Electromechanical Coupling Theory, Methodology and Applications for High-Performance Microwave Equipment

Electromechanical Coupling Theory, Methodology and Applications for High-Performance Microwave Equipment

Baoyan Duan and Shuxin Zhang
Xidian University, China

IEEE PRESS

WILEY

Published by John Wiley & Sons, Inc., Hoboken, New Jersey.
Published simultaneously in Canada.

For general information on our other products and services or for technical support, please contact our Customer Care Department within the United States at (800) 762-2974, outside the United States at (317) 572-3993 or fax (317) 572-4002.

Wiley also publishes its books in a variety of electronic formats. Some content that appears in print may not be available in electronic formats. For more information about Wiley products, visit our web site at www.wiley.com.

Library of Congress Cataloging-in-Publication Data Applied for

Hardback ISBN: 9781119904397

Cover Design: Wiley
Cover Image: © Solarseven/Shutterstock

Set in 9.5/12.5pt STIXTwoText by Straive, Chennai, India

Contents

About the Authors *xv*
Preface *xvii*

About the Authors

Baoyan Duan received the B. Eng., M. Eng., and Ph. D. degrees in Electromechanical Engineering from *XDU* in 1981, 1984, and 1989, respectively. From 1991 to 1994, he was a Post-doctoral Fellow at Liverpool University, U.K., and worked as Visiting Scientist at Cornell University, Ithaca, NY, in 2000. He is currently Academician of Chinese Academy of Engineering (*CAE*) and full Professor in the School of Electromechanical Engineering at *XDU* where he founded the Research Institute on Mechatronics. He was engaged as President of Xidian University (*XDU*), Xi'an, China, from 2002 to 2012.

Baoyan Duan has been dedicating himself to the research of electromechanical engineering and opened new areas of electromechanical coupling (EMC) theory in electromagnetic, structural deformation, and temperature fields of microwave electronic equipment (MEE). He has made known the influence mechanism (IM) of nonlinear mechanical parameters on electronic performance of MEE. He developed the integrated design methodology of MEE based on EMC and IM. The above academic achievements have been successfully applied in national major engineering projects such as the deep space exploration, the Shenzhou spacecraft, the Tiantong No.1- space deployable antenna, and so on.

As the Chief Design Engineer, he led the design team and was involved in implementing an innovative design called *optomechatronic design with electronic, mechanical, and optic technologies*, by which the millimeter dynamic-accuracy-positioning and ultra-light-weight (from 8,000 tons to 30 tons) were implemented for 500m diameter spherical radio telescope (*FAST500m*).

The telescope has been in operation since 2016 and many new planets have been observed for the first time. He was invited to provide a keynote speech on this achievement at EuCAP'2018 in London.

Baoyan Duan serves as the Chair of *Antenna Industry Alliance (AIA)* of China and Chair of *Electromechanical Engineering Society* of China. He is a Fellow of *International Engineering and Technology (IET)* and *Chinese Institute of Electronics (CIE)*, Members of Int. Society for *Structural and Multidisciplinary Optimization (SMO)*. He is engaged as the editor-in-chief of *Electromechanical Engineering* of China, the editor-in-deputy chief of *Chinese Journal of Electronics*, the Section editor-in-chief of *CAE* flagship magazine *ENGINEERING,* and the editor of 10 more academic journals.

Baoyan Duan has published 200 papers and six books and authorized 40 patents of invention. He has received, as the first author, the 1st prize of national award for science and technology progress (*STP*) of China 2020, and the 2nd prize of national award for *STP* of China three times (2004, 2008, and 2013). In 2009, he was selected as science Chinese person. In 2012, he was issued Hong Kong HLHL prize of *STP*. In 2017, he received the Award for Outstanding Scientific and Technological Achievement from *CAS* and the Golden Prize of Good Design of China. In 2018, he received the Life Achievement Award from the Asian Society of *SMO*.

CCTV (China Central TeleVision station) broadcasted a special program *DUAN Baoyan: Minor Discipline and Great Vision* in 2016.

Shuxin Zhang received the B. Eng. and Ph. D. degrees in mechanical engineering from Xidian University in 2006 and 2015, respectively. He is currently a Professor of Electromechanical Engineering in the Key Laboratory of Electronic Equipment Structure (Ministry of Education) in Xidian University.

Shuxin Zhang's research interests include the areas of integrated structural-electromagnetic analysis and optimization design of reflector antennas, and structural and multidisciplinary design of deployable antennas.

Shuxin Zhang has published 25 papers and authorized 15 patents of invention. He received the Excellent Doctoral Dissertation Nomination Prize bestowed by the Chinese Institute of Electronics in 2018.

Preface

Complex information and electronic equipment is a system that combines electromagnetism, mechanics, thermodynamics, and other disciplines. The successful realization of its electrical performance depends not only on the design level of various disciplines, but also on the organic combination of multiple disciplines. For example, the mechanical structure is not only is the carrier and guarantee for the realization of electrical performance, but also often restricts the realization of electrical performance. On the other hand, electrical performance also puts forward higher requirements on the mechanical structures. Therefore, mechanism and electrics are interrelated, interdependent, mutually influencing, and inseparable. Especially for complex electronic equipment with high frequency band, high gain, high density, miniaturization, fast response, and high pointing accuracy, the disciplines of mechanics and electrics show the characteristics of strong coupling. We call the mechanical and electrical interaction of electronic equipment the electromechanical coupling problem of electronic equipment.

The traditional mechanical and electrical separation design of complex electronic equipment leads to low performance, long cycle, high cost, and heavy structure in the development of electronic equipment. It has become a development bottleneck that has restricted the performance of electronic equipment in a long period of time and affected the development of equipment in the next generation. The detailed design procedure of the traditional electromechanical separation design is that the electrical designers put forward requirements for the mechanical mechanism design according to the requirements of the electrical performance, and the mission of the structural designers is to try to meet its detailed requirement. Due to the fact that the electrical designers do not have enough understanding of the difficulty of mechanical design and manufacturing, and that they leave enough space for the subsequent electric adjustment at the same time, the required accuracy is often difficult to meet, and sometimes it even exceeds the capability of mechanical manufacturing. On the other hand, due to the lack of understanding and mastery of electromagnetic knowledge for mechanical designers, the only way is to do everything possible to satisfy the accuracy. This traditional design introduces two problems. One is that the structure manufactured in accordance with the accuracy requirements cannot guarantee 100% of the electrical performance, and the other one is that the structure that does not meet the accuracy requirements often meets the electrical performances in contrary. This situation has existed for a long time and has not been fundamentally resolved.

It can be seen that the key to the design of electronic equipment is the study of electromechanical coupling. Specifically, it includes establishing a theoretical model of field coupling between the mechanical displacement field, electromagnetic field, and temperature field, exploring the influence mechanism of mechanical factors on electrical performance, proposing the electromechanical coupling test methods, evaluation methods, and multidisciplinary optimization design methods based on field coupling theory and influence mechanism, and developing a comprehensive design software platform that integrates electromagnetics, mechanics, and thermodynamics to fundamentally solve the long-standing bottleneck problem that restricts the improvement of electronic equipment performance and affects the development of electronic equipment in the next generation.

Based on this idea, the first author has been engaged in the scientific research and engineering application of electronic mechanical engineering since he was a master's student in the early 1980s, and is committed to the research and application of the interdisciplinary nature of electromechanical coupling technology of electronic equipment. After that, at the beginning of this century, on the basis of the long-term research work in the past, the national 973 project of "Research on Basic Problems of Electromechanical Coupling of Electronic Equipment" was further put forward and approved. With the joint efforts of a research team formed by technicians from universities, research institutes, and other units, substantial progress has been made in theory and technology, and a comprehensive design software platform has been developed, which has been applied in typical engineering cases and achieved satisfactory results.

The content of this book is a summary of past scientific research work. During the preparation of the manuscript, Congsi Wang, Fei Zheng, Jin Huang, Guangda Chen, Lihao Ping, and Yong Wang provided active help in different ways. Section 4.6 of the book is written by Jin Huang to complete the first manuscript, Chapters 5 and 6 are written by Guangda Chen to complete the first manuscript, and Chapter 8 is written by Congsi Wang to complete the first manuscript. In addition, the work of this book also includes the work of other comrades, such as Peng Li, Liwei Song, Hong Bao, Wei Wang, Shenghuai Zhou, Jinzhu Zhou, Hongbo Ma, Minbo Zhu, Fushun Zhang, Yongchang Jiao,

Xiansong Guo, Jundong Shi, Zhijian Yan, Changwu Xiong, and Zhenfang Shen. The authors would like to express heartfelt thanks to them.

The authors also express special thanks to the national 973 project expert team represented by academicians Lvqian Zhang, Guangyi Zhang, and Xixiang Zhang for their effective guidance in the progress of the project.

The first author is deeply grateful to his supervisor, Professor Shanghui Ye. Since 1982, he has studied and worked under the guidance of Professor Ye. He will never forget his mentor's guidance and help in many ways.

Due to the limited level of the authors and the rush of time, it is inevitable that there are some defects and deficiencies of one kind or another in the book. Readers are welcome to criticize and correct them.

This book can be used as a reference book for teachers, postgraduates, and senior undergraduates of colleges and universities, and engineering and technical personnel of research institutes.

1

Background of Electromechanical Coupling of Electronic Equipment

1.1 Introduction

Electronic equipment is a kind of special electromechanical equipment that targets the acquisition, transmission, and processing of electromagnetic signals or other electrical performance and adopts mechanical structures as the carrier.

Complex and high-performance electronic equipment is nothing but a system composed of multiple disciplines such as electronics, mechanical structures, and heat transfer. The successful acquisition of its performance depends not only on the design quality of each simple discipline but also, even more important, on the intersection and inosculation among them. This is because the mechanical structure not only guarantees the electrical performance as the carrier but also often restricts the realization and promotion of electrical performance. Therefore, it is necessary to study the electromechanical coupling of electronic equipment, toward establishing the electromechanical coupling theoretical model, making the influence mechanism of nonlinear structural factors on electromagnetic performance clear, and finally developing electromechanical coupling model and the influence mechanism-based system design theory and methodology.

Since the First Industrial Revolution, mechanical equipment has become the foundation of industrialization. Since the middle of the twentieth century, the rapid development of both electronic technology and computer technology has gradually made electronic equipment become an important equipment of modern industrial society. It has been widely applied in various fields such as land, sea, air, and space. With the development of science and technology, the demand of complex and high-performance electronic equipment is becoming more and more urgent.

The development of electronic equipment has passed through the age of electron tubes, the age of transistors, and the current era of integrated circuits. This is classified based on the devices. In terms of working frequency, it has

Electromechanical Coupling Theory, Methodology and Applications for High-Performance Microwave Equipment, First Edition. Baoyan Duan and Shuxin Zhang.

experienced the low-frequency era, the high-frequency era, and is moving toward ultrahigh frequency and terahertz (THz) frequency bands. Compared with the early electronic equipment, the prominent features of the modern electronic equipment are high frequency, high gain, large frequency width, high pointing accuracy, high density, and miniaturization, all of which put forward high requirements and greater challenges for the design and manufacturing of mechanical structural parts, some of which even exceed the limits of manufacturing ability. At this time, the electrical part and mechanical structure of the electronic equipment must be considered simultaneously and comprehensively, and the electromechanical coupling design is urgently needed [1, 2]. Unfortunately, the traditional design is independent between mechanical and electronic technologies, which leads to poor performance, long cycle, high cost, and unwieldy. It significantly restricts the improvement of the level of electronic equipment.

The way to solve the problem of electromechanical separation design is the electromechanical coupling design, which includes multifield coupling problem and the influence mechanism of electronic equipment. The research content of multifield coupling problem is to make the relationship between mechanics and electronics clear from the perspective of field coupling, and the task of influence mechanism is to find the influence mechanism of nonlinear structural factors on electrical properties. These two parts complement each other and are two manifestations of electromechanical coupling theory [3, 4].

The in-depth study of the electromechanical coupling of electronic equipment is the prerequisite for the development of high-performance electronic equipment. Early researches did not start from the level of electromechanical coupling. This was because the working frequency band was not high or the volume requirements were not harsh at that time. But now it is not the case, because mechanics and electronics are inseparable, and researches must be carried out from the system level of mechanics and electronics or mechanics, electronics, and heat transfer coupling to solve the problem thoroughly. There are two manifestations of electromechanical coupling, one is the form of field coupling, and the other is the influence mechanism. In engineering, the influence mechanism was understood in earlier times, and many researches were carried out. With the development of science and technology, the roles of different physical fields have been gradually recognized, so the electromechanical coupling problem of electronic equipment has begun to be investigated from the perspective of the field.

As for the concerned influence mechanism, the purpose is to understand, discover, and master the influence mechanism of mechanical structural factors on electrical performance and then provide empirical formulas, diagrams, and design specifications for guiding the electromechanical coupling design of electronic equipment. Mechanical structural factors include structural parameters and manufacturing tolerance. The field coupling theory is to provide a mathematical relationship between different physical fields, that is the mathematical model.

Generally speaking, coupled multifield problems (CMFP) refer to a physical phenomenon in which two or more physical fields affect each other through interaction. This phenomenon is widespread in objective world and practical engineering [5]. Common coupling problems include fluid–solid coupling, gas–solid–liquid coupling, structural–electromagnetic–thermal coupling, structural–optical coupling, acoustic–structural coupling, and electrostatic–structural coupling [6–15]. Through the analysis of some coupling phenomena appearing in actual engineering and the study of the inherent physical properties of each physical field itself, the mutual influence relationship between the physical fields can be initially determined (Figure 1.1). The inside of the circle in the figure is a physical field, the directed line segment indicates the interaction between the physical fields, and the text in the line segment indicates the physical quantity acting on it.

As for the CMFP appearing in electronic equipment, the coupling relationship among the physical fields can be simply expressed as follows. Taking the reflector antenna as an example, which is used in telecommunication, navigation, radar, radio telescope, and space deployable antenna, the reflector is used as the boundary of the electromagnetic field. When it is subjected to environmental loads (self-weight, wind, temperature, vibration, shock, etc.), the antenna reflector surface is deformed, which will change the boundary conditions of the electromagnetic field and affect the realization of electrical performance, such as gain degradation, increase in sidelobe level, and poor beam pointing accuracy. This is the first fact. In reverse, as the working frequency increases (the wavelength decrease), a tiny change in the electronic performance will give rise

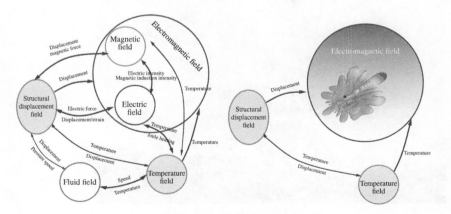

Figure 1.1 Coupling relationship diagram of each physics field.

to a big change in the antenna structure to guarantee this tiny change, which needs a large change to fit the performance. Meanwhile, as the working frequency becomes higher, the relationship between structure and electromagnetic performance becomes tier. From this viewpoint, the influence between antenna electronic and structural parameters is mutual and dual directional.

Furthermore, as the advent of the space era, the demand on spaceborne deployable antenna is urgent, of which the requirements are large diameter, high precision, light weight, and large ratio of the deployed size to furled size. To meet these requirements, the membrane antennas have been pushed in the research and application. Because of the low stiffness of membrane outside the surface, the electromagnetic pressure on membrane reflector from feed and space cannot be ignored. At this time, the influence between antenna electronic and structural parameters is mutual and dual directional [16].

1.2 Characteristics of Electronic Equipment

Modern electronic equipment have a wide variety of functions and appearances. The most typical electromagnetic signal receiving and transmitting equipment is the antenna. The reflector antenna is one of the most widely used form of antennas, such as the reflector antenna used in lunar and deep space exploration and the deployable antenna on communication satellites, as shown in Figures 1.2 and 1.3.

Figure 1.2 A 66 m-S/X-beam guide antenna for Mars detection.

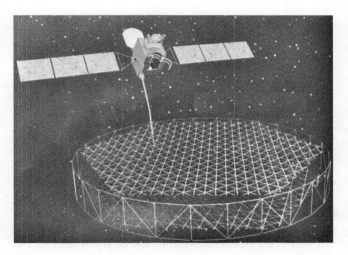

Figure 1.3 China "Tian Tong No. 1" spaceborne deployable antenna.

The earliest electromagnetic signal-processing equipment are radar transmitters, receivers, power amplifiers, and other equipment. Modern electronic technology usually uses the equipment as a functional module and then integrates them into one device, such as avionics, where each functional module is installed in the same chassis, and the lower part is a shared heat dissipation channel.

Modern electronic equipment have more powerful functions and more complex systems far beyond the scope of early electronic equipment. For example, in the Air Police 2000 early warning aircraft, as shown in Figure 1.4, not only the radar system and communication system on the aircraft are electronic equipment but

Figure 1.4 Early warning airplane – airborne phased array radar. Source: Twitter/ST.

Figure 1.5 The Yuan Wang ocean survey ship and antenna. Source: (a) baike.com.

also the entire early warning aircraft itself is an electronic equipment. Similar is the case for Yuan Wang ocean survey ship shown in Figure 1.5; regardless of its volume and size and how much pure is the mechanical equipment on it, considering the fact that the mission of the survey ship is to obtain and process the electromagnetic signals, it can also be called an electronic equipment.

The most notable feature of the electronic equipment is the integration with electromagnetic and mechanical technologies. Different from general mechanical equipment containing electrical components, the electronic equipment takes electrical performance as the main task of entire equipment, while the mechanical part is being subjected to electrical performance. That is to say, the mechanical part serves as the carrier and guarantees the realization of electrical performance.

Compared with general machines or traditional machines, the characteristics of electronic machines are embodied as follows: In terms of **purpose**, electronic machine pursues the electrical performance of electronic equipment systems, while conventional machine pursues their mechanical performance. In terms of the **realization means**, electronic machine is mainly realized by changing and optimizing mechanical structural parameters and processes, while traditional machine is mainly realized by adding electronic technology, optoelectronics, and other technologies.

Electronic equipment are one of the main research objects in the field of electromechanical engineering. Their essence is to study the crossover and integration of different disciplines, with a view to discovering theories, methods, and technical means to improve the performance of systems or equipment by studying the coupling problems between different physical quantities or physical fields. Electromechanical engineering mainly studies the mechanical design, structural design, and manufacturing of electronic equipment, information equipment, or electronic systems. Its characteristic lies in how to make the system or equipment meet the electrical performance requirements in the complex mechanical environment, electromagnetic environment, and thermal environment and retain high reliability.

1.3 Components of Electronic Equipment

1.3.1 Mechanical and Structural Part of Electronic Equipment

The mechanical and structural part of the electronic equipment includes two main aspects. One is the overall layout of the electronic equipment structure according to the working environment, the technical requirements, the overall conception of the system, and the design and planning of the subsystems. The other one is the design of mechanical parameters. For the equipment installed in the movable platform or transportation requirements, there should be sufficient strength and stiffness to resist a variety of environmental loads caused by material fatigue, structural resonance on the effects of electrical performance. If necessary, special vibration isolation and buffering measures have to be used too.

This is the most traditional area of structural design for electronic equipment and one of the earliest developed and most mature aspects of the design theory and methodology.

There are also requirements for electromagnetic compatibility (EMC) in the structural part. It also contains two aspects, one is the electrical performance and its component parts (devices) that can meet the requirements for the usage of electronic equipment, and the other one is the EMC requirements of the equipment as a whole component. The latter means that various measures are used to control electromagnetic interference from all aspects such as circuitry, structure, processing, and assembly and also to meet the cost requirements. It also contains two aspects of meaning, one is that the electronic equipment can resist external electromagnetic environment interference, and the other one is that the radiated electromagnetic waves from electronic equipment itself do not interfere with other electronic equipment.

For the structural design of electronic equipment, the key issue is to improve its electromagnetic shielding effectiveness, which is an indicator of EMC independent of the internal components and external environment. Whether the electronic equipment is shipboard, airborne, on land, or even spaceborne, there are specific requirements for electromagnetic shielding effectiveness.

Furthermore, there are also thermal control requirements on the structure of electronic equipment. The thermal design of electronic equipment is mainly for the electronic components and the whole mechanics for heat control. Unlike conventional machine and equipment, of which the heat is generated by energy or mechanical friction, the heat of electronic equipment is mainly generated from electronic devices. For modern high-density assembly equipment, the thermal capacity of the electronic component is much lower than the mechanical structure of the part; therefore, it requires more stringent thermal control measures, and the main purpose is to prevent the failure of electronic components due to high temperature.

This aspect reflects the difference between electronic equipment and other mechanical devices and has now been developed into a specific technology – thermal analysis and thermal control of electronic equipment, which encompasses three aspects of thermal analysis, thermal design, and thermal testing. Heat dissipation methods can be roughly divided into two types: air cooled and liquid cooled, both of which need to be selected according to the internal device power consumption and the external environment. Some equipment used in special environments may also require local heating, such as the space environment.

Other aspects of the structure that should be considered include the corrosion protection design, i.e. the selection of the appropriate protection measures and structure for the specific environment in which it is to be used. Ergonomic design, i.e. the requirement for ease of usage and maintenance by the operator. In addition, the design of the connections at the electrical contact points affects the reliability of the equipment and is increasingly attracting the attention of designers.

To sum up, the structural part of electronic equipment contains an extremely wide range of contents, involving mechanics, machinery, material science, thermal science, electromagnetism, environmental science, aesthetics, and many other fields, and is a marginal interdisciplinary discipline. The design should therefore also be carried out from a multidisciplinary perspective.

1.3.2 Electrical Part of Electronic Equipment

The electrical part can be divided into two categories. One is the circuit part, which includes the layout of various circuits, the design of printed boards, and even the design of a particular chip or electronic device.

The other category is the electromagnetic field section, mainly radar antennas, waveguides, etc., of which a certain distribution of the electromagnetic field in space is achieved by means of certain structures or electronic devices.

The electrical part inevitably has to be realized through the structural part, so the electrical part of the electronic equipment is inevitably influenced and constrained by the structural part.

1.4 On research of Electromechanical Coupling (EMC) of Electronic Equipment

1.4.1 Current Status of Research on Electromechanical Coupling of Electronic Equipment

The electromechanical coupling of electronic equipment is not only a wide range of fundamental theoretical issues but also one of the core technical issues that limit the electronic equipment for high performance, development cycle, and cost. It is to be regretted that few reports are known on the mechanism of mechanical, electrical, and thermal field coupling theory of electronic equipment and the mechanism of mechanical and structural factors affecting the electrical performance.

To begin with, there are some in-depth researches and applications in the component level or part [17–19]: For example, the influence of random wrinkle shapes on the transmission characteristics of electromagnetic waves [9], the electromechanical coupling and its influence coefficients in surface acoustic filters [10], the coupling between electrical and thermal fields in induction heating equipment [20], and the influence of mechanical structure manufacturing accuracy of phased array antennas on electrical performance [12]. In terms of the strong, but with very lower frequency, electromagnetic coupling of the motors, there are relatively many research references, such as the application of transient finite element method to calculate the coupling relationship between the outer coil and the rotor speed in the induction motor [13], etc.

Then, there are also sporadic reports on the system level of electronic equipment: For example, preliminary research on the integrated structural, electromagnetic, and thermal design of a spaceborne antenna reflector [14] and research on the analysis and layout of the thermal field of military electronic equipment [15].

One more aspect being worthwhile to mention is the researching activities from the Research Institute on Mechatronics of Xidian University, China. Since the early 1980s, teachers and students of the institute have successively carried out some in-depth researches [21–46]. For example, starting from the improvement of antenna gain, the optimal design of the shape for the secondary reflector of the dual reflector antenna is carried out. Starting from the concept of the antenna phase center, the problem of the field coupling relationship between the deformation field and the electromagnetic field of large reflector antenna is studied. A new mathematical relationship between structural displacement field and electromagnetic field was deduced mathematically. The influence of the supporting structure of the secondary reflector in dual reflector antenna on the electrical performance is studied. For different antenna attitudes, the influence of reflector surface deformation on antenna pattern (saying sidelobe) is analyzed. However, the abovementioned researches are limited to reflector antennas and are relatively preliminary.

Another aspect is tools; a number of commercial professional analysis software have emerged. These include ANSOFT for electromagnetic analysis, ANSYS for structural analysis, MARC, FLORMERICS, ALGOR, and SPECTRUM. These software can support the analysis of specific physical fields with good performance and later be able to add other physical field-processing functions. There are also software designs that are initially based on multifield problems, such as PHYSICA and COSMOL. There are also domain-specific multifield analysis software, such as Conventer Ware, IntelliCAD, and SOLIDIS for MEMS multifield simulation; NMSeses for fuel cell simulation; and CFD-ACE for 3D chip stack design. These tools are able to handle some multifield coupling problems but are limited to a limited number of products and lack wide applicability. They cannot establish a bridge between the electromagnetic field and the structural displacement field, and it is difficult to express the electromagnetic performance as a function of the structural design parameters. Unfortunately, simulation software for multifield coupling problems applicable to electronic equipment, especially the electronic equipment with high frequency and large wavelength dimension, has not been reported yet.

At the same time, the Research Institute on Mechatronics is also developing electronic equipment analysis and design software, including reflector antenna electromechanical comprehensive optimization design software, deployable antenna analysis and design software, and high-density assembly system analysis and design software. Prototype software systems have also been developed for the analysis and design of planar slotted and active phased array antennas. All the software are researched and developed from the perspective of multifield coupling, integrating the mathematical and quantitative relations of the coupling of each physical field into the software, ensuring the accuracy of the analysis, and improving the efficiency of the design.

1.4.2 The Development Trends of Electronic Equipment

With the rapid development of electronic technology, information technology, and even materials and processes, electronic equipment have also been developed rapidly, and its development trends are mainly reflected in the following aspects [47–61]:

1.4.2.1 High Frequency and High Gain

The resolution ability of electronic equipment, especially radar and antenna, is directly related to the wavelength, so the frequency of radar is getting higher and higher. In the early days, it was mainly meter-wave radar that achieved the basic function of ranging, but in modern times it has been developed into millimeter-wave high-resolution radar and is developing toward higher frequency. High gain is mainly used to increase the detection range of antenna. With the development of deep space exploration, the antennas with high gain are in great demand. The basic method to improve the antenna gain is to increase both the frequency and the antenna aperture. Modern large reflector antenna aperture has reached tens of meters or even hundreds of meters;

for instance, the aperture of the world's largest radio telescope named FAST is 500 m.

1.4.2.2 Broad Bandwidth, Multiband, and High Power

Other current and future requirements for electronic equipment are (i) broad bandwidth; for example, the frequency band of QTT 110 m full steerable telescope, under construction in Wulumuqi of the Northwest part of China, is between 300 MHz and 115 GHz, corresponding to wavelength from 100 cm to 2.6 mm or even wider; (ii) multiband, e.g. the requirement for the same antenna to operate on multiple bands; and (iii) high power, as in the case of equipment on satellites, where it is desirable to transmit as much power as possible while remaining the same size. These three requirements bring new problems and lead to great difficulties in the design and manufacturing of electronic equipment, for example more complex field coupling relations, higher processing accuracy requirements, the need for new materials, new structures, and new theories of exploration and research.

1.4.2.3 High Density and Miniaturization

Electronic devices are developing toward smaller, denser, and lower power consumption; for example electronic devices are being assembled more and more densely and from two- to three-dimensional assembly. Correspondingly, the size of the equipment is becoming smaller and smaller. For example, the size of a typical electronic equipment RF system has been reduced from $0.03\,\text{m}^3$ in 2000 to $0.01\,\text{m}^3$ in 2010s and is expected to reach $0.001\,\text{m}^3$ before long. The dramatic increase in density will lead to serious electromechanical coupling problems.

1.4.2.4 Fast Response and High Pointing Accuracy

The requirements for the mobility and responsiveness of the equipment are increasingly high, and they should be able to be precisely positioned while being fast tracking. For example, a shipboard radar antenna mount with a stabilized platform is required to be extremely fast, smooth at low speeds, and accurate in its positioning.

1.4.2.5 Good Environmental Adaptability

As human race explores space more and more fast, the environment in which electronic equipment service has become much hostile. Electronic equipment need to be able to work properly in a variety of abnormal environments such as space environments, alien environments, and deep sea. Military electronic equipment are even required to resist microwave, laser, and other new concepts of weapons attack or work under certain damage to preserve performance, while requiring the ability to combat strong electromagnetic interference and having a high degree of reliability. All of these require the electronic equipment to have a good environmental adaptability.

1.4.2.6 Integration

Integration is mainly for multidisciplinary, multifunctional, and high performance. For example, the design and manufacturing of these products require the integration with mechanical, electrical, and thermal disciplines. Secondly, the functional requirements are becoming increasingly demanding and need easy implementation of multifunctionality through modularization. Furthermore, the multifunctional requirements do not reduce the performance requirements in any way but rather place higher demands on performance.

1.4.2.7 Intelligence

Intelligence is another obvious trend in the development of electronic equipment. To realize their intelligence, the research of smart materials, smart control, and smart structures should be carried out first. In addition, artificial intelligence (AI), as enabling technology, could play an active role in the process of the intelligence of electronic equipment.

1.5 Problem of the Traditional Design Method of Electronic Equipment

1.5.1 Traditional Design Method and Problems with Electronic Equipment

For a long time, the development of electronic equipment is basically the electromechanical separation design. That is, the electrical designer first puts forward the requirements and manufacturing accuracy based on the electrical performance for mechanical or structural design. Then, it is up to the mechanical or structural designers to find ways to meet these requirements. Taking the development process of a high-density aviation chassis shown in Figure 1.6 as an example, its structural design must meet three requirements: structural rigidity, ventilation and heat dissipation, and electromagnetic screen efficiency. When the initial CAD model is provided according to the basic design requirements, the structural, thermal, and electromagnetic analysis models are established, respectively. The relevant analysis of the three disciplines is performed, and their respective preliminary design schemes are provided. The three schemes are bound to be conflicted with each other. The chief designer needs to coordinate the three schemes based on experience and finally come up with a feasible

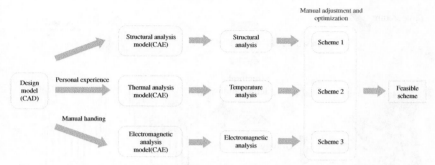

Figure 1.6 Diagram of the traditional electromechanical separation design of the chassis.

scheme. In the early days of electronic equipment development, this was a feasible design method. However, with the development of technology and the improvement of requirements, this method of electromechanical separation has become increasingly difficult to meet the design requirements of modern electronic equipment. It has become a bottleneck that restricts the improvement of equipment performance and affects the development of equipment in the next generation.

1.5.2 The Electromechanical Coupling Problem of Electronic Equipment and Its Solution

The mutual influences and restrictions of electromagnetic and mechanical structures in electronic equipment are inseparable. With the development of researching activities, it is discovered that the mechanical structure and electromagnetic interaction in electronic equipment are in the form of a field. Taking the reflector antenna shown in Figure 1.7 as an example, the antenna reflector is a boundary of the electromagnetic field. Under the load of gravity, wind, snow, etc., the antenna reflector surface will be deformed, and it will inevitably deviate from the theoretical design surface. This deformation will affect the antenna's gain (efficiency), sidelobe, and other electrical properties. Moreover, as the working frequency increases, this influence relationship becomes more prominent. Another example is the field coupling problems among the structural displacement field, electromagnetic field, and temperature field of missile-borne electronic equipment, which will seriously affect the guidance accuracy of the missile. Furthermore, the radar antenna in the airborne, shipborne, and other moving environments directly affects its pointing accuracy. It can be noticed that the field coupling relationship among the structural displacement field, electromagnetic field, and temperature field in electronic equipment is universal.

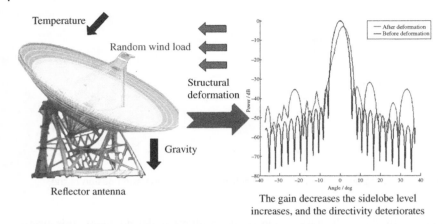

Figure 1.7 The influence of reflector antenna structure on electrical performance.

1.6 Main Science and Technology Respects of Design for Electronic Equipment

1.6.1 Holism of Electronic Equipment System Design

Electronic equipment is a system that combines multiple disciplines. Its composition mainly includes the mechanical structure part and the electromagnetic part. Its performance includes electrical performance and mechanical performance. These two parts are not separated and independent but are interacted with each other and closely related. Therefore, it should be considered as a whole system. In addition to the electrical part, the mechanical and structural part should be considered when studying the electrical performance, and the electrical performance should also be paid attention to consider the structural design. During the design, both the structural factors and electrical factors should be considered at the same time. Only by studying electronic equipment as a whole unit can we design high-performance electronic equipment that meet the requirements.

1.6.2 Electromechanical Coupling Theory of Electronic Equipment

The electromechanical coupling theory of electronic equipment includes two aspects: one is the field coupling theory, and the other is the influence mechanism.

The so-called field coupling theory refers to the mutual influence among the electromagnetic field, structural displacement field, and the temperature field existing in electronic equipment. For example, in a reflector antenna, under the external effects of its self-weight, wind, etc., the structure will be deformed, which will cause a change in the shape of the antenna reflector. It is ultimately reflected in the electrical performance of the antenna. Therefore, it is necessary to study the two-field coupling relationship between the electromagnetic field and structural displacement field. As another example, in addition to electromagnetic field and structural displacement field, high-density enclosures also need to consider thermal issues, that is a three-field coupling relationship among electromagnetic field, displacement field, and temperature field needs to be established. The purpose of establishing the field coupling relationship is to reveal the internal relationship between the fields, express the electrical performance as a function of the structural design variables, and lay a solid theoretical foundation for the electromechanical coupling design.

The so-called influence mechanism refers to the discovery of the influence law of mechanical structure factors on electrical performance. Because there is a type of problem, that is the electrical properties are difficult to be expressed in the form of fields, such as manufacturing nonparallelism, nonperpendicularity, and roughness, these quantities are mostly related to the processing route with great randomness and are generally collectively referred to as manufacturing accuracy. Material properties, stiffeners, bosses, grooves, etc. are issues that need to be considered in the structural design and are generally referred to as structural parameters collectively. The abovementioned manufacturing accuracy and structural parameters can be collectively referred to as structural factors. There are two ways to study the influence mechanism of structural factors on electrical properties, one is induction, and the other is deduction. The induction is based on the existing massive data, using the support vector regression method to get the law that satisfies the input and output samples. The deductive method is based on the problem of establishing a simulation model. Whether it is induction or deduction, the final influence mechanism is a design specification in the form of empirical formulas and diagrams.

1.6.3 Test and Evaluation Methods of Electronic Equipment

The testing of complex electronic equipment is one of the foundations of the application of electromechanical coupling theory and methods. The basic amount of test data must be obtained, otherwise it cannot be carried out. The method of evaluation of complex electronic equipment is to evaluate the correctness and effectiveness of the field coupling theory and influence mechanism. Specifically, it includes two aspects. One is the electromechanical coupling test method, referring to the modeling and calculation of the coupling degree of test factors, the test technology, and the construction of the test database. The other is the comprehensive evaluation method of electromechanical coupling model and influence mechanism, including evaluation system construction and calculation method.

The test method of electromechanical coupling is aimed at high-performance electronic equipment and to study economic and effective test strategies, methods, and techniques of electromechanical coupling based on the idea of comprehensive integration. Through the application of new test strategies, methods, and technologies to improve the efficiency and technical level of electronic equipment testing, the establishment of integrated test systems and the testing of typical cases can be guided. The specific content includes researches on new technologies and test strategies for the measurement of mechanical quantities and electrical parameters and integrated electromechanical test systems based on data integration.

The evaluation method of electromechanical coupling is to investigate the establishment and evaluation method of electromechanical coupling evaluation system based on electrical performance and manufacturability in electronic equipment. It needs to study the evaluation methods of typical electronic equipment from a theoretical perspective and provide theoretical and methodological guidance for the design and manufacturing of electronic equipment. The evaluation of electromechanical coupling based on electrical performance is to judge the improvement of the overall electrical performance index of equipment due to the application of electromechanical coupling theory and influence mechanism under the condition that the manufacturing cost remains unchanged. Based on the evaluation of manufacturability, the manufacturing cost is taken as the specific evaluation target, and the cost changes brought by the impact of the debugging cycle and manufacturing accuracy requirements are studied with the help of electromechanical coupling theory and influence mechanism with the same electrical performance. The specific research content includes the establishment method of the evaluation system and the comprehensive evaluation method of electromechanical coupling.

1.6.4 Environmental Adaptability (Thermal, Vibration, and EMC) and Reliability of Electronic Equipment

How to prevent failure of electronic equipment and make the electronic equipment work reliably under strong vibration and impact are the topics of electronic equipment design. The temperature of electronic equipment is controlled in a harsh environment so that the temperature of electronic components and devices does not exceed the allowable value. The abilities of electronic equipment are to resist external electromagnetic interference and avoid their own electromagnetic pollution to the environment. In addition, there are antimoisture, antimildew, antisalt spray corrosion, antiatom, antibiochemical weapons and so on. This involves many different disciplines such as mechanics, heat transfer, electromagnetic field theory, environmental science, chemistry, and materials science. Moreover, various protective measures must be integrated and unified in the design.

How to correctly and effectively connect, assemble, and lay out thousands of electronic components and devices to form a complete mechanics or system is also a topic of designing electronic equipment. In the assembly process, the mutual influence of various electronic components and devices must be considered internally, and the influence of various environmental factors must be considered externally. Ultimately, high reliability, easy maintenance, and easy operation must be ensured. At present, electronic assembly has developed to the level of surface mount and microassembly. In microassembly, the circuit, structure, and process are inseparable. In addition, ergonomic considerations cannot be ignored for the purpose of ensuring operators to work efficiently.

1.6.5 Special Electronic Equipment

Certain special-purpose electronic equipment must adopt special structures, such as super large or super flexible structures. Typical representatives are a new generation of large radio telescopes and large spaceborne deployable antennas.

(1) Spaceborne deployable antennas

Due to the limitation of rocket launch capacity, the antenna on the satellite generally adopts a stowed volume, which is folded when launching and deployed after into the normal space orbit. At the same time, in order to reduce weight, the cable net structure or membrane structure instead of the commonly used metal reflector is utilized as the antenna reflective surface. Both the cable net structure and membrane structure are all flexible structures. On the other hand, in order to make the antenna have enough gain, the diameter of the antenna becomes larger and larger, which increases the difficulty in design and manufacturing. Therefore, the large-scale satellite deployable antenna is not only a flexible structure but also a super large structure.

(2) Large radio telescopes

The plan for the new generation of large radio telescope (LT) [61, 62] was proposed by world astronomers at the Kyoto Conference in Japan in 1993 and had received unanimous support from 10 countries including China, the United States, Britain, and France. FAST 500 m super large radio telescope situated in Guizhou Province, Southwest part of China, named China Space Eye (CSE), as shown in Figure 1.8, became the world largest radio telescope.

In the innovation design with integration with optic, mechanical, and electronic technologies, the original rigidity support and servo system is replaced with six

(a)   (b)

Figure 1.8   FAST 500 m largest radio telescope. (a) Bird's view of the FAST telescope.
(b) Feed cabin and cable system.

long suspending cables, so that about 10 000 tons self-weight of the feed cabin
support structural system is decreased to 30 tons. Each suspension cable is driven
by a servo system, and the six servo systems are coordinated and controlled by
a central control computer. At the same time, the idea of active main reflector
surface is adopted. That is, the 500-m spherical reflector surface is spliced by
thousands of small regular triangular plates, and the back of each small regular
triangular has three actuators to make it possible to change the orientation as
required, so that the illuminated part can be turned into a parabolic surface in
real time. The radio telescope is not only super large (500 m) in size but also
super flexible structures with six long-span suspension cables. In order to get
the millimeter positioning accuracy of the feed, a fine Stewart tuning platform is
installed within the cabin. Moreover, the A/B adjusting axis is used between the
cabin and Stewart platform to decrease the pressure on the Stewart platform.

1.6.6   Electromechanical Coupling Design of Electronic Equipment

As typical electromechanical devices, electronic equipment have structural
characteristics and electromagnetic characteristics that influence and restrict
each other. Only from the perspective of electromechanical coupling and interdisciplinary, the high-performance electronic equipment could be designed.
Therefore, coupling design has become the preferred method for structural design
of typical electronic equipment, including antenna parts, servo systems, and
high-density chassis.

1.6.6.1   Electromechanical Coupling Design of Antennas
As early as the 1960s, researchers had noticed that the structural and electrical
performances of antennas were intrinsically linked. Antenna structural designers also began to investigate the relationship between structural parameters and

the electrical performance of antennas. During the times, a series of researching results were obtained, and the theory of best-fit paraboloid and conformal design was introduced and summarized in several monographs on antennas published in the 1980s. The contents of these monographs include antenna electromagnetic design, structural design, and servo system design of antenna mounts, covering almost all the fields involved in radar antennas, which became the classic work on reflector antenna design in China and has been used until now.

Since the beginning of the new century, through the long-term research, researchers have discovered that the structural parameters and electrical performance of the antenna interact through the form of fields on the basis of the best-fit parabolic, conformal design theory, and electromechanical integration design ideas. The structural displacement field and electromagnetic field are coupled. Therefore, the electromechanical coupling theory of the antenna was proposed, and the optimization design of electromechanical coupling was carried out. At the same time, the theory of electromechanical coupling is extended to other antenna forms, such as planar slotted array antennas and active phased array antennas. The design idea of electromechanical coupling is also applied to the analysis and design of the antenna servo system.

However, for optimum design by electromechanical coupling of reflector antennas, the consideration of electrical performance is still based on the Ruze formula, and the electromechanical coupling formula is not really applied yet. The surface accuracy of the reflector surface is calculated through structural deformation, and then the gain loss is calculated from the surface accuracy according to the Ruze formula. On the one hand, a single main surface accuracy cannot fully reflect the structural deformation of the antenna, and the position error of the subreflector and feed is not considered. On the other hand, gain is just one electrical performance pursued by an antenna. In most of the cases, performances such as sidelobe level and pointing accuracy are much more important. The current electromechanical coupling design lacks all these features.

1.6.6.2 Integrated Design of Radar Antenna Servo System

The electromechanical system is composed of two subsystems: mechanism (or structure) and control. The integrated design of these two systems is very necessary. The research on the integrated design of structure and control began in the 1980s. Scholars have conducted fruitful researches, mainly focusing on the following three aspects: The first is the integrated design of the space system structure and control. By changing the structural parameters (such as cross-sectional area), controller position, and gain factor, optimization methods such as nested optimization, genetic algorithm, and response gradient are used to achieve the dual goals of light weight and low control energy consumption simultaneously. The second is the integrated design of the structure and control of the DC motor. The structure and control factors are connected through the transfer function. Based on the configuration and motion demand constraints, by changing the number of coil turns, the motor's full-load rated power, and PID control parameters, the quality of the motor, control errors, and energy consumption are minimized. The first two categories are integrated optimization design problems for invariable structures and are not suitable for optimization problems of variable structures (or mechanisms). The third is the integrated design of structure and control in the mechanism system. The design is performed from the first control-oriented structural design, then the integrated design method of the optimal controller, to the subsequent two-stage nested optimization design method of structure and control iteration, and finally to the comprehensive optimization method of optimizing both the parameters of structure and control of the mechanism at the same time. According to the structural invariance of a single-rotating beam during motion, the literature combines frequency domain transformation of dynamic equations and control theory to obtain a control system and then optimizes the structural parameters to achieve the goal of minimizing quality and control energy consumption. However, this method is not suitable for complex mechanisms with variable structural characteristics and many design variables (resulting in a huge amount of calculation). For example, for a rigid four-bar linkage mechanism, it can achieve the goal of the smallest dynamic tracking error and the least energy consumption by optimizing parameters of the cross-sectional area of the rod, the weight, and the control. Unfortunately, the influence of the flexibility of the rods on the movement of the mechanism is not considered, and the natural frequency and stability constraints of the mechanism are not considered either.

The performance of the radar antenna's pointing accuracy and fast response depends on the design level of its servo system. The design of the servo system includes two parts, i.e. structural and control design. Structural design will affect the realization of the control performance. For example, the realization of servo control bandwidth depends on the natural frequency of the structure. Conversely, the control will affect the structural design. For instance, the size of the driving force in the servo system will affect the design of the antenna base structure. Therefore, in order to achieve the objectives of "seeing accurate" and "seeing clear," the structure and control have to be integrated in the design procedure.

However, the traditional design of radar antenna servo systems is separated between the structural and control design; that is, the mechanical structure and control system are designed independently and then tuned to meet the required specifications. In fact, the structure and control of the radar antenna servo system are coupled with each other; especially in high-performance tracking, these two parts are closely coupled. If the characteristics of the servo structure are not fully considered in the control design, the servo tracking performance will be reduced or even unable to achieve the required performances. On the other hand, if the control part is not fully considered in the structural design, the optimal design cannot be obtained or even fail to meet the performance requirements. This separation design approach leads to long product development cycles, high costs, poor performance, and bulky structures. Traditional design methods have constrained the development of high-performance radars. For this reason, it is necessary to integrate the structure and control together for comprehensive optimization design to achieve the best overall performance.

1.6.6.3 Coupling Design of High-Density Chassis

The three main aspects of high-density chassis design, namely, the structural rigidity, ventilation and hear transfer heat dissipation, and EMC, are contradictory to each other. It is embodied as follows: One is the contradiction between quality and rigidity. The structural strength is required to be as high as possible; namely, it is required to be able to work normally under the impact and vibration of various working conditions, and the working environment requires small size and light weight, especially for airborne and ammunition-borne equipment. The other is the contradiction between EMC efficiency and ventilation and heat dissipation. Larger holes are good for heat dissipation but not good for electromagnetic shielding. Excessive temperature will affect the working efficiency of electronic devices. Modern electronic equipment chassis is asked to meet the three requirements of structural rigidity, EMC, and ventilation and heat dissipation at the same time.

The conventional chassis design is to separately consider the above three requirements of structural rigidity, EMC, ventilation and heat dissipation, respectively, and provide separated design schemes. Since the starting points and purposes are different, there will be conflicts between the design schemes. Therefore, the chief designer is required to make a balance and trade-offs based on experience and arrives at a feasible design scheme. This is an effective design method when the early requirements are not strict in all aspects. However, with the increasing requirements of various aspects, this design method of structural–electromagnetic–thermal separation becomes much difficult to meet the requirements of various aspects at the same time. As a result, it is necessary to carry out research from the perspective of multifield coupling, establish a multifield coupling model, and then propose a multidisciplinary optimization model based on the multifield coupling model for structural–electromagnetic–thermal coupling design.

1.7 Mechatronics Marching Toward Coupling Between Mechanical and Electronic Technologies

High-precision and high-performance complex equipment have been extensively applied in the important areas such as national defense development, national economy, and high and new technologies, of which the design and manufacturing capabilities are significant embodiments of national science and technology level and strength. Complex equipment mainly consist of two categories: one is the precision mechanical equipment focused on mechanical performance, and the electrical performance is subject to mechanical performance, including manufacturing equipment such as large numerical control machines, machining center, etc., and the industrial major equipment such as armaments, chemical engineering, shipping, agriculture, energy, digging, and tunneling. The electronic technologies are mainly employed to reform, arm, or improve the mechanical performance of the traditional equipment. The other one is the electronic equipment focused on electronic performance, and the mechanical performance is subject to electronic performance, such as radar, communication, computer, navigation, antenna, and radio telescopes. Electronic equipment, of which the mechanical structures are mainly to guarantee the electromagnetic performance, are widely used and play irreplaceable roles in the key areas such as land, sea, air, and space.

Generally speaking, both categories of equipment belong to complex equipment with integration of mechanical and electronic technologies and are typical cases of key applications of mechatronic technologies. The concept of mechatronics, originated in the 1970s, is nothing but a combination of two words, i.e. mechanical and electronics, and reflects the connotation evolution and development trend of continuous fusion of mechanical and electromagnetic (electrical) technologies.

With the development of mechatronic technologies, new concepts, such as mechanical–electrical–liquid integration, fluid–sound–gas integration, and biologic–electromagnetic integration, arise in succession. These concepts, although with different names, essentially belong to the scope of mechatronics and study the interrelation among different physical systems or fields, thus improving the overall performance of systems or devices.

From the viewpoint of mechatronic design, high-performance complex equipment have experienced three different phases, i.e. Independent between Mechanical and Electronic Technologies (IMET), Syntheses between Mechanical and Electronic Technologies (SMET), and Coupling between Mechanical and Electronic Technologies (CMET). The development of high-precision and high-performance electronic equipment shows highlighted features of these three phases.

IMET means that the mechanical design and electromagnetic design of electronic equipment are separated and independent of each other, whereas the information can be conveyed and shared on/off-line. The mechanical and electromagnetic designs are independently implemented in each domain. On the boundary or within each domain, the sharing and effective transmission of information can be realized. Typical examples include the structure–electromagnetics of reflector antennas and the temperature–structure–electromagnetics of active phased array antennas.

It should be pointed out that from the design level, such kind of information sharing is still IMET, and thus the inherent problems of traditional independent design also exist. There are two most remarkable problems: one is that the requirements on the mechanical design and manufacturing precision proposed by electromagnetic designers are usually too high to be implemented, and the mechanical designers, deficient in deep comprehension of electromagnetic knowledge, can just try their best to satisfy the requirements. It is sort of blindly. The other one is that, in practical engineering, there exist strange phenomena that the manufactured products, with full effort, sometimes unsatisfied the electronic performance, whereas the electrical performance of some products with manufacturing precision below the required level may meet the requirements. Therefore, in practical engineering, the method of copies has to be utilized and the choice has to be made after electronic test and measure. These two longtime-existed problems have led to the low performance, long period, high cost, and heavy structure in electronic equipment development, thus being an open bottleneck that seriously restricts the improvement of electronic equipment performance and affects the development of next-generation equipment.

With the constant rise in electronic equipment working frequency, the interaction between mechanics and electromagnetics tends to be more remarkable, and the IMET design is confronted by more problems and more serious contradictions. Thus, the design concept of SMET arises. SMET is a relatively high level of mechatronics and is a large stride advance of IMET due to two factors: one is the establishment of the integrated design mathematical model that can consider simultaneously mechanical, electromagnetic, and thermal performance, etc. and effectively eliminate certain deficiencies and shortcomings; the other one is the establishment of integrated finite element analysis models; for example, in the analysis of high-density chassis and cabinet, the numerical analysis model of electromagnetics, structure, and temperature within the same geometrical space can be shared.

From the beginning of this century, electronic equipment take on three new developing trends such as high frequency and high gain, high density and miniaturization, and fast response and high pointing accuracy and strong coupling features of the relation between mechanical and electromagnetic technologies. In this way, mechatronics strides into the new phase of CMET.

CMET is the further step toward ideal mechatronics than SMET, which includes two main characteristics: one is that, mechanical, electromagnetic, and thermal automatic numerical analysis and simulation can be realized, and the completeness, accuracy, and reliability can be guaranteed during the information transmission among different disciplines. The other one is that the coupling theoretical model of multiphysics system is deduced mathematically based on physical quantity coupling, and the influence mechanism of nonlinear mechanical factors on electronic performance is revealed. In this way, the design becomes the CMET design based on the coupling theoretical model and the influence mechanism. Therefore, CMET is inherently different from SMET and is essentially advanced over SMET.

From IMET and SMET to CMET, mechatronic technologies show distinct generation evolution, which provide theoretical and key technology support for high-level equipment design and manufacturing. The future development of complex equipment manufacturing tends to be a deep fusion of multiphysical field, multimedium, multidimension, and multielement, and mechanics, electrical science, electronics, electromagnetics, optics, and thermology will be focused together. Huge system, extremalization, and precision treatment will be the new tendency, and greater challenges will be confronted by the design and manufacturing technologies with CMET as a breakthrough.

With the fast development of new-generation electronic technologies, information technologies, and materials and processing techniques, the development of future high-performance electronic equipment will take on two extreme features: one is extreme frequency, such as the extremely low-frequency band for submarine communication, or millimeter wave, submillimeter wave, and even terahertz wave for spaceborne microwave radiation antenna applications. The other one is extreme environments, such as the North and South Poles of the Earth, deep space and near space (20–100 km from the ground), deep sea, etc. These factors present unprecedented challenges for CMET theories and technologies, and it is urgent to carry out the following studies.

Firstly, the establishment of electromechanical coupling theoretical model of electromagnetic, structural displacement, and temperature fields. Due to the relation of interaction and inter-restriction among them, it is necessary to reveal the influence and coupling mechanism; ascertain the coupling mechanism of multifields, multidomain, multidimension and multimedium, and the influence mechanism of multiworking condition and multifactors; and represent them as quantitative mathematical relations.

Secondly, the nonlinear mechanical factors (structural parameters and manufacturing precision) and material properties of electronic equipment have apparent effect on the electromagnetic performance. It is urgent to explore the influence rules of these nonlinear factors on electromagnetic performance and further reveal the influence mechanism (IM) on electromagnetic performance.

Thirdly, CMET design methods. It is necessary to synthetically analyze the characteristics of coupling theoretical model and IM and thus propose theories and methods of electronic equipment CMET design. This involves the independent analysis model of mechanics, electromagnetics and thermology, and the treatment of difficult points such as the slide of numerical analysis meshes among them.

Fourthly, the mathematical representation and measurement of coupling degree. In theory, any coupling could be measurable. To deeply explore the coupling behavior among multiphysical systems, it is necessary to explore a general mathematical representation method for measurement coupling and further deduce the mathematical expression for quantitative calculation of coupling degree.

Finally, deep fusion in applications. CMET technologies not only exist in almost all electromechanical equipment but also play an important role in the transformation and upgrading of high-level equipment manufacturing. CMET technologies are the common key technologies for iteration development and applicable to many major industries in the development of equipment manufacturing, thus penetrating the whole historical process of industrialization and informatization. With the arrival of new scientific and technological revolution and industrial reform, especially with the appearance of intelligent manufacturing featured as digitization, networking, and intellectualization, the deep fusion of industry and information technologies is imperative. Such fusion shows itself as the application of CMET theories in the level of theories and technologies, and it can be thus seen that CMET is of profound significance and promising future.

References

1 Duan, B. and Song, L. (2008). On coupled multi-field problems in electronic equipments. *Electro-Mechanical Engineering* 24 (3): 1–7+46. (in Chinese).

2 Felippa, C.A., Park, K., and Farhat, C. (2001). Partitioned analysis of coupled mechanical systems. *Computer Methods in Applied Mechanics and Engineering* 190 (24-25): 3247–3270.

3 Zhong, J. (2007). *Coupling Design Theory and Method of Complex Electromechanical Systems*. Beijing: China Machine Press.

4 Zhong, J. and Chen, X. (1999). Coupling and decoupling design for complex electromechanical system. *China Mechanical Engineering* 10 (9): 10–12. (in Chinese).

5 Song, S. (2007). *Research and Application of Collaborative Solution Method for Multiphysics Problems*. Huazhong University of Science and Technology (in Chinese).

6 Song, Z., Zhang, W., and Shi, A. (2010). *Fundamentals of Fluid-Structure Coupling and Its Application*. Harbin: Harbin Institute of Technology Press (in Chinese).

7 Kamakoti, R. and Wei, S. (2004). Fluid–structure interaction for aeroelastic applications. *Progress in Aerospace Sciences* 4: 535–558.

8 Sun, P., Yang, D., and Chen, Y. (2007). *Introduction to Coupling Models for Multiphysics and Numerical Simulations*. Beijing: Science and technology of China press (in Chinese).

9 Amari, S., Vahldiech, R., and Bornemann, J. (1999). Analysis of propagation in corrugated waveguides with arbitrary corrugation profile. In: *IEEE AP-S International Symposium*. Vig, Orlando, USA, July, 1999, 290–293. IEEE.

10 Yaralioglu, G.G., Ergun, A.S., Bagram, B. et al. (2006). Calculation and measurement of electromechanical coupling coefficient of capacitive micromachined ultrasonic transducers. *IEEE Transactions on Ultrasonic, Ferroelectrics, and Frequency Control* 50 (4): 449–456.

11 Dughiero, F. (1998). Numerical and experimental analysis of an elector-thermal coupled problem for transverse flux induction heating equipment. *IEEE Transactions on Magnetics* 34 (5): 35–46.

12 Wang, H.S.C. (1992). Performance of phased-array antennas with mechanical errors. *IEEE Transactions on Aerospace & Electronic Systems* 28 (2): 535–545.

13 Pham, T.H., Wending, P.F., Salon, S.J. et al. (1999). Transient finite element analysis of an induction motor with external circuit connections and electromechanical coupling. *IEEE Transactions on Energy Conversion* 14 (4): 1407–1412.

14 Adelman, H.M. and Padula, S.L. (1986). Integrated thermal-structure-electromagnetic design optimization of large space antenna reflectors. NASA Technical Memorandum 87713.

15 Price, D.C. (2003). A review of selected thermal management solutions for military electronic systems. *IEEE Transactions on Components and Packaging Technologies* 26 (1): 26–39.

16 Zhang, X.H., Zhang, S.X., Cheng, Z.A. et al. (2017). Structural-electromagnetic bidirectional coupling analysis of space large film reflector antennas. *Acta Astronautica* 139: 502–511.

17 China Defense Science and Technology Information Center (CDSTIC). http://210.79.226.16:81 (last accessed 1 December, 2005).

18 NASA Science and Technology Report. http://www.sti.nasa.gov (last accessed 1 December, 2005).

19 Swedish Defense Research Service. http://www.foi.se (last accessed 1 December, 2005).

20 Dughiero, F. (1998). Numerical and experimental analysis of an electro-thermal coupled problem for transverse flux induction heating equipment. *IEEE Transactions on Magnetics* 34 (5): 35–46.

21 Guohua, X. and Shi, H. (1984). Pre-optimized design of the surface shape of the dual-reflector antennas. *Journal of Northwest Telecommunications Engineering Institute* 4: 3–19. (in Chinese).

22 Liu, J. (1988). Electromechanical synthetic optimization of antenna's subreflector support structures. *Journal of China Institute of Communication* 9 (6): 62–65. (in Chinese).

23 Qi, Y. and Hongshi, W. (1992). The hybrid phase center of a reflector antenna feed source. *Acta Electronica Sinica* 20 (9): 22–26. (in Chinese).

24 Li, Y. (1985). Research on mechatronics of sidelobes of dual-reflector antenna. *Journal of Northwest Telecommunications Engineering Institute* (in Chinese).

25 Wang, W. (1988). Research on mechatronics design technology of reflector antenna system—the influence of structural parameters on cross-polarization. *Journal of Xidian University* (in Chinese).

26 Guohua, X., Qi, Y., Duan, B. et al. (1990). A study of the phase center of the antenna feeder for a deformed reflector. *Journal of Xidian University* 17 (4): 63–70. (in Chinese).

27 Wang, W. and Guohua, X. (1994). The effect of reflector surface distortion on the antenna radiation pattern. *Acta Electronice Sinica* 22 (12): 46–49. (in Chinese).

28 Qi, Y. (1989). Systematic technology on the integration of electronics-mechanics of reflector antennas. *Journal of Xidian University* (in Chinese).

29 Guohua, X. and Wang, J. (1990). Analysis of mechatronics system of reflector antenna. *Communication & Measurement & Control* 3: 43–49. (in Chinese).

30 Qi, Y. (1991). Mechatronics design of reflector antenna. *Chinese Journal of Radio Science* 6 (1): 150–152. (in Chinese).

31 Guohua, X. and Shao, Z. (1996). Design ideas of antenna electromechanical system engineering. *Electro-Mechanical Engineering* 5: 30–34. (in Chinese).

32 Liu, G. (2004). Antenna mechanical-electronic integral analysis. *Radio Communications Technology* 30 (4): 25–26. (in Chinese).

33 Ye, S. and Chen, S. (1982). Guided weight criterion method for optimal design of antenna structure. *Journal of Northwest Telecommunications Engineering Institute* 1: (in Chinese).

34 Duan, B. (1986). A comprehensive method of optimal design of antenna structure. *Journal of Northwest Telecommunications Engineering Institute* 3 (3): 57–65. (in Chinese).

35 Duan, B. and Ye, S. (1985). Geometrically optimised design of antenna structures with discrete variables. *Journal of Northwest Telecommunications Engineering Institute.* 3: (in Chinese).

36 Duan, B. (1989). Integrated topology, shape and electromechanical optimization of antenna structures. *Journal of Xidian University* (in Chinese).

37 Duan, B. (1991). Study of geometric representation and its sensitivity analysis in the optimal design of continuum shapes. *Journal of Xidian University* 18 (4): (in Chinese).

38 Duan, B. (1998). *Analysis, Optimization and Measurement of Antenna Structures.* Xian: Xidian University Press (in Chinese).

39 Duan, B.Y., Qiu, Y.H., and Xu, G.H. (1994). Study on optimization of mechanical and electronic synthesis for the antenna structural system. *Mechatronics* 4 (6): 553–564.

40 Duan, B. (1999). Review of Multidisciplinary Optimization of Antenna Structures in China. *Electronics Machinery Engineering* 79 (3): 2–6. (in Chinese).

41 Duan, B.Y. and Wang, C.S. (2009). Reflector antenna distortion using MEFCM. *IEEE Transactions on Antennas Propagation* 57 (10): 3409–3413.

42 Wang, C.S., Duan, B.Y., and Qiu, Y.Y. (2007). On distorted surface analysis and multidisciplinary structural optimization of large reflector antennas. *International Journal of Structural and Multidisciplinary Optimization* 33 (6): 519–528.

43 Ma, H., Duan, B., and Wang, C. (2009). Deformed reflector antenna with random factors and integrated design with mechanical and electronic syntheses. *Chinese Journal of Radio Science* 24 (6): 1065–1070. (in Chinese).

44 Song, L.W. (2010). Performance of planar slotted waveguide arrays with surface distortion. *Progress in Electromagnetics Research Symposium*, March 22–26, 2010, Xi'an, China.

45 Wang, C.S., Duan, B.Y., Zhang, F.S. et al. (2010). Coupled structural-electromagnetic-thermal modelling and analysis of active phased array antennas. *IET Microwaves, Antennas & Propagation* 4 (2): 247–257.

46 Wang, C.S., Duan, B.Y., Zhang, F.S. et al. (2009). Analysis of performance of active phased array antennas with distorted plane error. *International Journal of Electronics* 96 (5): 549–559.

47 Kamal, Y.T. (1996). Modeling, design and control integration: a necessary step in Mechatronics. *IEEE/ASME Transactions on Mechatronics* 1 (1): 29–37.

48 Onoda, J. and Haftka, R.T. (1987). A approach to structure/control simultaneous optimization for large flexible spacecraft. *AIAA* 25: 1133–1138.

49 Rao, S.S. (1988). Combined structural and control optimization of flexible structures. *Engineering optimization* 13: 1–16.

50 Yamakawa, H. (1989). A unified method for combined structural and control optimization of nonlinear mechanical and structural systems. *International Journal of Computer Aided Optimum Design of Structures* 287–298.

51 Iwadare, M., Kajiwara, I., Tsuchiya, R. et al. (2004). Integrated actuator/control design of smart pantograph mechanism for vibration suppression. In: *Proceedings of the 2004 IEEE International Conference on Control Applications*, 1717–1722. Taipei: IEEE.

52 Reyer, J.A. and Fathy, H. (2001). Comparison of combined embodiment design of control optimization strategies using optimality conditions. *ASME Design Engineering Technical Conferences*, Paper DAC-21119, September 9–12, 2001.

53 Reyer, J.A. and Papalambros, P.Y. (2002). Combined optimal design and control with application to an electric DC motor. *Transactions of the ASME, International Journal of Mechanical design* 124 (6): 183–191.

54 Semba, T., Huang, F., and White, M.T. (2003). Integrated servo/mechanical design of HDD actuators and bandwidth estimation. *IEEE Transactions on Magnetics* 39: 2588–2590.

55 Zhang, W.J., Li, Q., and Guo, L.S. (1999). Integrated design of mechanical structure and control algorithm for a programmable four-bar linkage. *IEEE/ASME Transactions on Mechatronics* 4 (4): 354–362.

56 Wu, F.X., Zhang, W.J., Li, Q. et al. (2002). Integrated design and PD control of high-speed closed-loop mechanisms. *Journal of Dynamic Systems, Measurement, and Control* 124: 522–528.

57 Ouyang, P.R., Li, Q., and Zhang, W.J. (2003). Integrated design of robotic mechanisms for force balancing and trajectory tracking. *Mechatronics* 13: 887–905.

58 Ke, F. and Mills, J.K. (2005). A convex approach solving simultaneous mechanical structure and control system design problems with multiple closed-loop performance specifications. *Journal of Dynamic Systems, Measurement and Control, Transactions of the ASME* 127 (3): 57–68.

59 Zhu, D.L., Jiang, T., Wei, J.H. et al. (2006). Integrated optimal model of structure and control of the single arm manipulator. *Journal of Beijing Institute of Technology (English Edition)* 15 (9): 278–282.

60 Yan, H.S. and Yan, G.J. (2009). Integrated control and mechanism design for the variable input-speed servo four-bar linkages. *Mechatronics* 19 (3): 274–285.

61 Duan, B. (2005). *Flexible antenna structure analysis, optimization and precision control*. Beijing: Science Press (in Chinese).

62 Duan, B. (2021). Mechatronics: toward electromechanical coupling technology. *Science & Technology Review* 39 (5): 1–2. (in Chinese).

2

Fundamental of Establishing Multifield Coupling Theoretical Model of Electronic Equipment

2.1 Introduction

In high-performance electronic equipment, structural displacement field, electromagnetic field, and temperature field are more closely related to one another (see Figure 2.1), and their influence on equipment performance is more prominent [1, 2]. Therefore, the in-depth study of the electromechanical–thermal field coupling relationship for the development of high-performance electronic equipment has important theoretical significance and engineering application value.

For typical cases such as large reflector antennas, planar slotted array antennas, active phased array antenna, and high-density assembly system, a field coupling model among structural displacement field, electromagnetic field, and temperature field should be established, where for reflector antennas and planar slotted array antenna, they are two typical kinds of microwave equipment with prominent two-field coupling problems in electronic equipment. To solve this coupling problem, a coupling model of structural displacement field and electromagnetic field of reflector antennas and planar slotted array antennas (electromechanical coupling model) should be established, and its correctness is verified by examples. Based on the electromechanical coupling model of reflector antennas and planar slotted array antennas, the optimization design model from structural design variables (type, topology, shape, size) to electrical performance is established, which lays a theoretical foundation for the electromechanical coupling optimization design of typical cases.

For another kind of electronic equipment, such as active phased array antennas and high-density assembly system, the field coupling relationship is more complicated as the coupling of structural displacement field, electromagnetic field, and temperature field. To this end, it is necessary to first investigate the theoretical problem of the three-field coupling of the structural displacement field, electromagnetic field, and temperature field, to establish the theoretical model of the electromechanical–thermal three-field coupling of active phased array antennas

Electromechanical Coupling Theory, Methodology and Applications for High-Performance Microwave Equipment,
First Edition. Baoyan Duan and Shuxin Zhang.
© 2023 The Institute of Electrical and Electronics Engineers, Inc. Published 2023 by John Wiley & Sons, Inc.

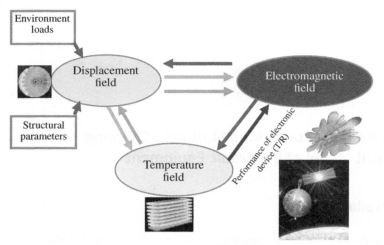

High-power microwave transmission

Figure 2.1 Electromechanical–thermal field coupling diagram. Source: RISH, Kyoto University.

and high-density assembly systems and to verify the correctness through several typical cases.

2.2 Mathematical Description of Electromagnetic (EM), Structural Deformation (*S*), and Temperature (*T*) Fields

To facilitate the discussion, the following differential equations describing several physical fields are first introduced, and on this basis, the quantitative coupling relationships between the physical fields can be derived by analyzing the physical nature and influence relationships of each physical field.

2.2.1 Electromagnetic Field

The electromagnetic field follows Maxwell's equations, which can be described as follows:

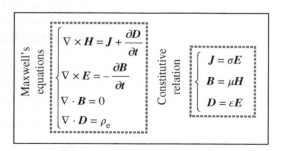

$$\text{Maxwell's equations} \begin{cases} \nabla \times H = J + \dfrac{\partial D}{\partial t} \\[2mm] \nabla \times E = -\dfrac{\partial B}{\partial t} \\[2mm] \nabla \cdot B = 0 \\[2mm] \nabla \cdot D = \rho_e \end{cases} \quad \text{Constitutive relation} \begin{cases} J = \sigma E \\[2mm] B = \mu H \\[2mm] D = \varepsilon E \end{cases}$$

$$(2.1)$$

In the system of equations (2.1), ∇ is a differential operator, H is the magnetic field intensity, E is the electric field intensity, B is the magnetic flux density, D is the electric flux density, J is the free current density, ρ_e is the free charge density, and σ, μ, and ε are the electrical conductivity, magnetic permeability, and dielectric constant of the material, respectively.

The first and second equations on the left-hand side of the system of equations (2.1) describe the relationship between the electric and magnetic fields vividly, indicating that electric current and changing electric field will generate the magnetic field, and the changing magnetic field will generate the electric field. The third and fourth equations provide the properties of magnetic field and electric field, respectively. The third equation represents the continuity of magnetic flux, that is there is no free magnetic charge. The fourth equation indicates that the charge generates the magnetic field, and free charge density is the source of electric field divergence.

The characteristic equation of the electric field, i.e. the fluctuation equation, can be written as

$$\nabla^2 E + k^2 \nabla E = 0 \tag{2.2}$$

where $k = \omega\sqrt{\varepsilon\mu}$, $\omega = 2\pi f$.

The current in a medium with a dielectric constant of ε satisfies the following equation:

$$I = \varepsilon \frac{\partial E}{\partial t} \tag{2.3}$$

2.2.2 Structural Displacement Field

The displacements and stresses of an elastic structure satisfy the following set of differential equations of elastic mechanics – Lame's equations, which are shown below:

$$\text{Lame's equations}\begin{cases} (\lambda^S + G)\dfrac{\partial\Theta}{\partial x} + G\nabla^2 u^S = \rho^S \dfrac{\partial^2 u^S}{\partial t^2} \\[2mm] (\lambda^S + G)\dfrac{\partial\Theta}{\partial y} + G\nabla^2 v^S = \rho^S \dfrac{\partial^2 v^S}{\partial t^2} \\[2mm] (\lambda^S + G)\dfrac{\partial\Theta}{\partial z} + G\nabla^2 w^S = \rho^S \dfrac{\partial^2 w^S}{\partial t^2} \end{cases} \quad \begin{aligned} \lambda^S &= \dfrac{E^S v}{(1+2v)\,(1-v)} \\[4mm] G &= \dfrac{E^S}{2(1+v)} \end{aligned} \tag{2.4}$$

In Eq. (2.2), the function Θ is related to the three components of stress as $\Theta = \sigma_x + \sigma_y + \sigma_z$; u^s, v^s, and w^s are the three components of displacement; ρ^s, E, and v represent the material density, modulus of elasticity, and Poisson's ratio, respectively; and t is the time parameter.

If the region to be solved is dissected as a finite element mesh and the variation principle is applied, a second-order differential equation for performing numerical analysis is obtained as

$$M\ddot{\delta} + C\dot{\delta} + K\delta = F \tag{2.5}$$

where M, C, and K are the overall mass, damping, and stiffness matrices of the structure, respectively; F is the array of external loads at the nodes; and $\ddot{\delta}$, $\dot{\delta}$, and δ are the array of accelerations, velocities, and displacements at the nodes.

2.2.3 Temperature Field

In general, the temperature field exists in three modes of heat transfer: conduction, convection, and radiation, and a set of descriptive equations for the different heat-transfer modes needs to be provided, specifically.

Heat conduction equation	$\rho^S c \dfrac{\partial T}{\partial t} = \tau^S \nabla^2 T + q_v$
Heat convection	$h_x = \dfrac{\tau^T}{T_{wx} - T_f}\left(\dfrac{\partial T}{\partial y}\right)_{y=0}$
Continuity equation	$\dfrac{\partial \rho^T}{\partial t} + \nabla\cdot\left(\rho^T U\right) = 0$
Energy conservation equation	$\dfrac{\partial T}{\partial t} + U\cdot\nabla T = \dfrac{\tau^T}{\rho^T c_p}\nabla^2 T$

Momentum conservation equation

$$\rho^T\left(\frac{\partial u^T}{\partial t} + U\cdot\nabla u^T\right) = f_x^T - \frac{\partial p}{\partial x} + \eta\nabla^2 u^T$$

$$\rho^T\left(\frac{\partial v^T}{\partial t} + U\cdot\nabla v^T\right) = f_y^T - \frac{\partial p}{\partial y} + \eta\nabla^2 v^T$$

$$\rho^T\left(\frac{\partial w^T}{\partial t} + U\cdot\nabla w^T\right) = f_z^T - \frac{\partial p}{\partial z} + \eta\nabla^2 w^T$$

$$\tag{2.6}$$

Thermal radiation equation $\Phi_r = \varepsilon^T \sigma^T A^T T^4$ (2.7) where ρ^T, c, T, t, τ^T, and q_v are the heat flow density, specific heat capacity, temperature, time, thermal conductivity, and heat source, respectively; ε^T, Φ_r, σ^T, and A^T are the emissivity, radiant heat, Stephan–Boltzmann constant, and radiate area, respectively; u^T, v^T, and w^T are the velocity components, along three axes, of the liquid flow vector U. The superscript T indicates the quantity in heat transfer, while the superscript S in Eq. (2.2) indicates the quantity of structure. T_{wx}, T_f, and h_x are the temperature at any x, the solid–liquid boundary, and the boundary layer heat-transfer coefficient, respectively; c_p, p, and η are the constant pressure specific heat capacity, pressure, and kinetic viscosity coefficients, respectively; and f_x, f_y, and f_z are the source terms of the liquid momentum conservation equation, respectively. Note that the viscous dissipation term S_T, which should be at the right-hand side of the equation, is not considered in the energy conservation equation.

In both the above equations and the following discussion, the following operators are used, which are briefly described below:

$$\text{Dispersion} \quad \nabla \bullet A = \frac{\partial}{\partial x} + \frac{\partial}{\partial y} + \frac{\partial}{\partial z} \quad \text{or} \quad divA = \frac{\partial}{\partial x} + \frac{\partial}{\partial y} + \frac{\partial}{\partial z}$$

$$\text{Rotation} \quad \nabla \times A = \begin{bmatrix} i & j & k \\ \dfrac{\partial}{\partial x} & \dfrac{\partial}{\partial y} & \dfrac{\partial}{\partial z} \\ A_x & A_y & A_z \end{bmatrix}$$

Laplace operator $\nabla^2 = \dfrac{\partial^2}{\partial x^2} + \dfrac{\partial^2}{\partial y^2} + \dfrac{\partial^2}{\partial z^2}$ is equivalent to performing the scatter $\nabla \bullet (\nabla)$ operation on the gradient, i.e.

For the scalar function f, the Laplace equation is $\nabla^2 f = \nabla \bullet (\nabla f) = 0$.
For the vector function $\vec{f}$, the Laplace equation is $\nabla^2 \vec{f} = \nabla \times (\nabla \times \vec{f}) = \nabla(\nabla \bullet \vec{f}) - \Delta \vec{f} = 0$

2.3 Consideration of Establishing Multifield Coupling Model

The establishment of mathematical model is the basis of the coupling of multiple fields problem (CMFP). Fluid–structure interaction is a typical example among CMFP. The mathematical model of CMFP in general sense can be derived from the description of the fluid–structure coupling mathematical model.

The mathematical model of fluid–structure interaction problem has been provided in many researches [3–5]. After giving the descriptive equations of fluid and solid, boundary and initial conditions, respectively, a mathematical model

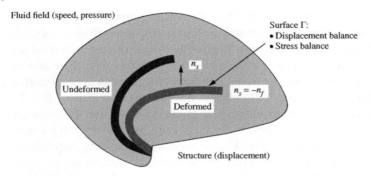

Figure 2.2 Geometric model of fluid–structure interaction.

of the fluid–solid interaction problem can be established by determining the equilibrium problem of the common boundary between fluid and solid. Being with the coupling relationship analysis of fluid and structure, it can be known that there are two kinds of equilibrium conditions on the common boundary: displacement equilibrium condition and stress equilibrium condition (Figure 2.2).

The equilibrium condition of the common boundary can be expressed as

$$u_s \cdot n_s + u_f \cdot n_f = 0 \quad \text{on} \quad \Gamma \quad \text{displacement equilibrium condition} \quad (2.8)$$

$$\sigma_s \cdot n_s + \sigma_f \cdot n_f = 0 \quad \text{on} \quad \Gamma \quad \text{stress equilibrium condition} \quad (2.9)$$

where u_s, u_f, σ_s, σ_f, n_s, and n_f are the displacement vector, stress vector, and unit vector normal to the boundary of solids and fluids, respectively, and Γ represents the boundary shared by solid and fluid simultaneously.

By combining the dynamic differential Eqs. (2.4) and (2.5) for mechanical structure; the three conservation equations for the mass (continuum), energy, and momentum of fluid in descriptive heat-transfer Eq. (2.6); and the common boundary equilibrium conditions (2.8) and (2.9), the set of equations describing the fluid–solid coupling problem can be obtained, i.e. the mathematical model of fluid–solid coupling problem. Equations derived here are only a general description of the coupling relationship. In view of the specific problem, it is necessary to deduce the expression form of coupling relation required by description equation of solid and fluid with mathematical physics method. Details about this will not be repeated here.

The mathematical description of the fluid–solid coupling problem tells us that there is a transformation of coupling information among physical fields with coupling relations, i.e. there is an interaction among the field variables in the mutually coupled physical fields, and the mathematical model of CMFP in this general sense can be expressed as follows:

Supposing that the field A can be mathematically described as

$$f_A(x_A, x_{A \to B}, x_{B \to A}) = 0 \quad \text{in} \quad \Omega_A \tag{2.10}$$

The field B can be mathematically described as

$$f_B(y_B, y_{B \to A}, y_{A \to B}) = 0 \quad \text{in} \quad \Omega_B \tag{2.11}$$

The influence of the field A on the field B can be described as

$$C_{A \to B}(x_{A \to B}, y_{A \to B}) = 0 \quad \text{in} \quad \Omega_{A \to B} \tag{2.12}$$

The influence of the field B on the field A can be described as

$$C_{B \to A}(x_{B \to A}, y_{B \to A}) = 0 \quad \text{in} \quad \Omega_{B \to A} \tag{2.13}$$

where f_A and f_B are differential operators, x_A and y_B are independent variables of field A and B, $x_{A \to B}$ and $x_{B \to A}$ are variables in field A that affect field B and the variables in field A that are influenced from field B, respectively, $y_{B \to A}$ and $y_{A \to B}$ are variables in field B that affect field A and the variables in field B that are influenced from field A, respectively, $C_{A \to B}$ and $C_{B \to A}$ are differential or algebraic operators, and Ω_A, Ω_B, $\Omega_{A \to B}$, and $\Omega_{B \to A}$ are the corresponding domains.

Equations (2.10)–(2.13) are the set of coupling equations for the CMFP. Only the mathematical description of the CMFP in a general sense is given here. For the specific CMFP, it is necessary to determine the various variables in the coupling model and the mode and form of the coupling action to establish a specific mathematical description of the specific CMFP.

As for the theoretical model of the coupled electromagnetic, displacement, and temperature fields of electronic equipment, a mathematical description of the coupled electromechanical–thermal fields of electronic equipment in the general sense can be obtained by introducing the correlation among the fields on the joint basis of the Maxwell's equations (2.1), the elasticity equations (2.4), and the heat-transfer equations (2.6).

The mathematical model of the physical field is partial differential equation or equations [6, 7] with time and space coordinates acting as the independent variables. The numerical methods widely used in engineering field are divided into finite element method, boundary element method, finite difference method, and finite volume method, according to different meshing methods. Although the meshing methods of physical fields are basically different, the basic framework of numerical analysis process is similar.

It can be seen from Figure 2.3 that although the discretization of each physical field in the region, the discretization of the description equation, and the discretization of the initial and boundary conditions are different, the solution of each physical field problem is transformed into the solution of algebraic equations. Therefore, it is possible to establish a unified field model in the sense

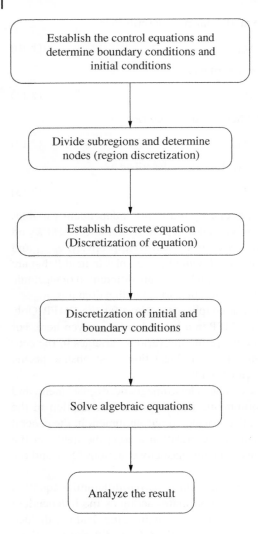

Figure 2.3 Numerical analysis flow of each physical field.

of regional dispersion and describe the discrete equation of each physical field using the unified field model. Each physical field can be constructed in the form of its unified field model. The expression of coupling relation is related to the unified field model of each physical field through discretization, and the mathematical model under the meaning of CMFP discretization is established. The key to achieving this is to establish a unified field model which is suitable for each physical field. To build a unified field model, what is necessary to be deeply investigated is not only the physical essence of each physical field but also the common problems among each physical field.

References

1 Bao-Yan, D. and Li-Wei, S. (2008). On coupled multi-field problems in electronic equipments. *Electro-Mechanical Engineering* 24 (3): 1–7+46. (in Chinese).

2 Duan, B.-Y. and Song, L.-W. (2007). Modeling and solving of multiple physical fields for electronic equipment. *Celebration of the 50th anniversary of Building Chinese Society of Mechanics and the Conference of Chinese Society of Mechanics.* (in Chinese).

3 Benney, R. and Stein, K. (1996). A computational fluid–structure interaction model for parachute inflation. *Journal of Aircraft* 33: 730–736.

4 Boujot, J. (1987). Mathematical formulation of fluid-structure interaction problems. *Mathematical Modeling and Numerical Analysis* 21: 239–260.

5 Moller, H. and Lund, E. (2000). Shape sensitivity analysis of strongly coupled fluid–structure interaction problems. In: *AIAA 2000-4823, 8th AIAA/USAF/NASA/ISSMO Symposium on Multidisciplinary Analysis and Optimization*, Long Beach, CA, vol. 9, 6–8. AIAA.

6 Hermann, G.M., Niekamap, R., and Steindorf, J. (2006). Algorithms for strong coupling procedures. *Computer Methods in Applied Mechanics and Engineering* 195: 2028–2049.

7 Ladeveze, P., Neron, D., and Schrefler, B.A. (2005). A computational strategy suitable for multiphysics problems. *International Conference on Computational Methods for Coupled Problems in Science and Engineering*.

3

Multifield Coupling Models of Four Kinds of Typical Electronic Equipment

3.1 Introduction

As far as the typical high-performance microwave equipment is concerned, antennas and high-density assembly systems are widely applied in telecommunication, navigation, radar, radio astronomy, and so on. Correspondingly, the following four kinds of microwave equipment are discussed in detail.

3.2 Reflector Antennas

For the ideal centered reflector antenna shown in Figure 3.1 (in the diagram, xoy is the aperture plane, f is the focal length, and the diameter of the reflector surface is $2a$), the electromagnetic field vector distribution of the aperture surface, radiated from feed via reflector, can be obtained. The far-field pattern of antenna can be known through the amplitude distribution of the aperture field (the phase is the same).

$$E(\theta, \phi) = \iint_A E_0(\rho', \phi') e^{jk\rho' \sin\theta \cos(\phi - \phi')} \rho' d\rho' d\phi' \tag{3.1}$$

$$E_0(\rho', \phi') = \frac{f(\xi, \phi')}{r_0} \tag{3.2}$$

where (θ, ϕ) is the far-field observation direction, A is the aperture surface area projected from the reflector onto the plane xoy, $f(\xi, \phi')$ is the primary pattern of the feed, and $k = 2\pi/\lambda$ is the wave constant associated with wavelength λ. In terms of the dual reflector antennas usually used in the engineering, the feed and sub-reflector can be equivalent to a feed on the virtual focus of the subreflector by equivalent feed method.

The analysis shows that the main factors related to the structure of the electromagnetic field include reflector panels, feed position, and orientation. However,

Electromechanical Coupling Theory, Methodology and Applications for High-Performance Microwave Equipment, First Edition. Baoyan Duan and Shuxin Zhang.

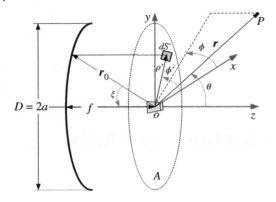

Figure 3.1 Geometric relation of centered reflector antenna.

the external loads will cause the variant of the structural displacement field such as reflector deformation, feed position deviation, and attitude deflection. Therefore, the main factors affecting the electrical performance are reflector surface errors, feed position, and pointing errors. In the following section, the relationship between various errors and the amplitude and phase distribution of the electromagnetic field on the aperture surface is discussed, aiming to provide the field coupling model of the displacement field and electromagnetic field in the case of various errors in the reflector antenna [1–11].

3.2.1 Influence of Main Reflector Deformation

The main reflector surface error consists of two parts, namely random error and systematic error. The random error mainly occurs in the manufacturing and assembling of panels, backup structure, and the center body. There are three ways to describe the random error. First, according to the specific processing technology, the mean and variance of the random distribution are calculated from the numerous data, with the assumption of a reasonable distribution. The specific distribution function can be obtained. Second, the distribution function is generated randomly by computer based on the mean and variance values. Third, the fractal function is used to describe the amplitude, frequency, and roughness caused by machining directly, and then the corresponding distribution function can be generated. In either way, the distribution function is superimposed on the systematic error, which is then involved in the electrical performance calculation as a unified error.

Systematic error refers to the antenna surface deformation caused by external loads such as self-weight, environmental temperature, and wind, and it is a deterministic error. The systematic error can be obtained by finite element analysis of the structure.

Figure 3.2 Reflector surface
error diagram.

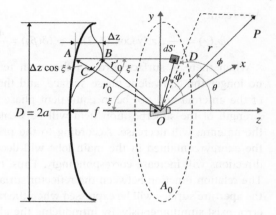

Since the reflector is supposed to be located in the far-field region of the feed, and thus, under the condition of small amplitude deformation of the main reflector, the influence made by the electromagnetic field vector distribution sent from the feed to the aperture plane can be ignored, and only the phase error of the aperture plane is considered. The main reflector surface error can be expressed as the axial error or the normal error. To facilitate the discussion, the axial error is used. As can be seen from Figure 3.2, when there is no error on a certain part of the reflective surface, the electromagnetic wave will go along the path of $OCABD$, and when there is error Δz, the electromagnetic wave will go along the path of OBD. The difference between these two paths is CBD (C is the point where the perpendicular line of B intersects OA), so the optic path difference between the two path is

$$\tilde{\Delta} = AB + AC = \Delta z(1 + \cos\xi) = 2\Delta z \cos^2(\xi/2) \tag{3.3}$$

Thus, the aperture phase error under the influence of the reflector surface error can be obtained as

$$\varphi = 2\pi\frac{\tilde{\Delta}}{\lambda} = k\tilde{\Delta} = \frac{4\pi}{\lambda}\Delta z \cos^2(\xi/2) \tag{3.4}$$

The profile error of the main reflector surface contains both random and systematic errors, i.e.

$$\Delta z = \Delta z_r(\gamma) + \Delta z_s(\delta(\beta)) \tag{3.5}$$

where γ is the random error caused during manufacturing, assembly, etc., $\delta(\beta)$ is the structural displacement, and β is the structural design variable, including structural size, shape, topology, type, and other parameters.

Thus, Eq. (3.4) becomes

$$\varphi = \frac{4\pi}{\lambda}(\Delta z_r(\gamma) + \Delta z_s(\delta(\beta)))\cos^2(\xi/2) = \varphi_r(\gamma) + \varphi_s(\delta(\beta)) \tag{3.6}$$

As a result,

$$\varphi_r(\gamma) = \frac{4\pi}{\lambda} \Delta z_r(\gamma) \cos^2(\xi/2), \quad \varphi_s(\delta(\beta)) = \frac{4\pi}{\lambda} \Delta z_s(\delta(\beta)) \cos^2(\xi/2) \tag{3.7}$$

When there is surface error in the main reflector, the aperture surface is no longer the equivalent phase surface, and the radiation field along the axis of the antenna will not be in equivalent phase with each other, and then the strength of the superposition field will be weakened. As a result, the gain of the antenna will decrease. According to the principle of energy conservation, the energy contained in the main lobe will decrease, and the energy in other directions will increase correspondingly. Thus, the sidelobe level will increase. The relation function between the reflector surface error and the phase error of the aperture surface will be expressed when the random error and the systematic error exist simultaneously. By introducing the phase error information into the electromagnetic field analysis model, a mathematical model made by the main reflector surface error on the electrical performance of the reflector antenna can be obtained. Thus, Eq. (3.1) will change into

$$E(\theta, \phi) = \iint_A E_0(\rho', \phi') \cdot \exp j[k\rho' \sin\theta \cos(\phi - \phi')]$$
$$\cdot \exp j[\varphi_s(\delta(\beta)) + \varphi_r(\gamma)] \rho' d\rho' d\phi' \tag{3.8}$$

3.2.2 Influence of the Feed Position Error

Under the influence of external loads, the surface of the reflector antenna will be deformed, and the reflector's feed position deviation and attitude deflection will be introduced. Therefore, the influence of the feed position and pointing error on electrical performance should also be considered.

The concept of feed position errors means position changes in the position of the feed phase center. When the position error of the feed phase center is small, the effect of the position error on the amplitude of the electromagnetic field on the aperture surface can be ignored, and only the phase error on the aperture surface is considered. Supposing that the feed position error is $\vec{d}$, then as shown in Figure 3.3, the vector can be obtained as

$$\vec{r}_0' = \vec{r}_0 - \vec{d}(\delta(\beta)) \approx \vec{r}_0 - \hat{r}_0 \cdot \vec{d}(\delta(\beta)) \tag{3.9}$$

in which $\hat{r}_0$ is the unit vector along $\vec{r}_0$ direction.

Thus, the aperture phase error under the influence of the feed position error could be written as

$$\varphi_f(\delta(\beta)) = k\hat{r}_0 \cdot \vec{d}(\delta(\beta)) \tag{3.10}$$

The error of the feed phase center also causes the phase error of the aperture plane. When the feed has an error along the axis (longitudinal error), the aperture

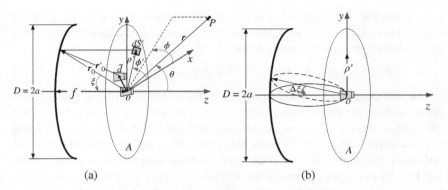

Figure 3.3 The feed error. (a) Feed position error. (b) Feed pointing error.

phase error is symmetric, which is similar to the square phase error. At this time, the maximum radiation direction of the far field remains unchanged, where the gain decreases, the sidelobe level increases, and the width of the main lobe increases. If the feed is of an error along the direction which is perpendicular to the axis (transverse error), the aperture phase error is close to the linear phase error, and the maximum radiation direction of the main lobe of the antenna pattern will deviate from the axis by a certain angle. At this time, the pattern becomes asymmetric, and the sidelobe level near the axis will increase significantly, while the sidelobe level on the other side will decrease. The width of the main lobe will not change much, and the gain loss is small. By introducing the phase error information into the electromagnetic field analysis model, a mathematical model of the influence of the feed position error on the electrical performance of the reflector antenna can be obtained. That is, the Eq. (3.1) will be changed as follows:

$$E(\theta, \phi) = \iint_A E_0(\rho', \phi') \cdot \exp j[k\rho' \sin \theta \cos (\phi - \phi')]$$
$$\cdot \exp j[\varphi_f(\delta(\beta))]\rho' d\rho' d\phi' \tag{3.11}$$

3.2.3 Effect of Feed Pointing Error

Feed pointing error can be interpreted as a shift in the primary pattern of the feed. When there is a pointing error $\Delta\xi$ between the feed and the direction of the negative z-axis, the new pointing angle can be found according to the geometric relationship of the pointing error of the feed as shown in Figure 3.3b,

$$\xi' = \xi - \Delta\xi(\delta(\beta)) \tag{3.12}$$

Due to the displacement field of the antenna structure, the pattern $f(\xi, \phi')$ of the feed will also have pointing error in the direction ϕ', i.e.

$$\tilde{\phi}' = \phi' - \Delta\phi'(\delta(\beta)) \tag{3.13}$$

By replacing the variables ξ, ϕ' in the feed pattern $f(\xi, \phi')$ with ξ', $\tilde{\phi}'$ in Eqs. (3.12) and (3.13), the feed pattern being subjected to the feed pointing error can be described as the following form:

$$f_0(\xi', \tilde{\phi}') = f_0(\xi - \Delta\xi(\delta(\beta)), \phi' - \Delta\phi'(\delta(\beta))) \tag{3.14}$$

The feed pointing error will bring about the amplitude error of the aperture plane, and the maximum radiation direction of the antenna will not change, but the sidelobe level will increase. As it is described, the feed position error will bring the phase error of the aperture plane field distribution, and the feed pointing error will cause the amplitude error of the aperture plane field distribution. Due to the fact that the two types of errors in the feed have different influences on the behaviors of electromagnetic field, a model of the relationship between the feed error and the electromagnetic field can be obtained by superposition of their influences. That is, under the influence of the feed source error, the normalized field distribution on the aperture plane can be mathematically expressed as

$$E_0 = \frac{f_0(\xi - \Delta\xi(\delta(\beta)), \phi' - \Delta\phi'(\delta(\beta)))}{r_0} \exp j[\varphi_f(\delta(\beta))] \tag{3.15}$$

By introducing this feed error into the analytical model, a mathematical model of the effect of feed error on the electrical performance of the reflector antennas can be obtained. Eq. (3.1) can be written as

$$E(\theta, \phi) = \iint_A \frac{f_0(\xi - \Delta\xi(\delta(\beta)), \phi' - \Delta\phi'(\delta(\beta)))}{r_0}$$
$$\cdot \exp j[k\rho' \sin\theta \cos(\phi - \phi')] \cdot \exp j[\varphi_f(\delta(\beta))]\rho' d\rho' d\phi' \tag{3.16}$$

3.2.4 Electromechanical Two-field Coupling Model

In practical engineering, external loads such as self-weight, wind, and temperature will deform the reflective surface with distortion Δz_s, feed position deformation $\vec{d}$, and attitude errors $\Delta\xi$, $\Delta\phi'$, which will ultimately lead to a reduction in the electrical performance of the reflector antennas. To reflect the effect of these errors, the following electromechanical two-field coupling model of the effect of main reflector surface systematic error, feed position error, and pointing error on the electromagnetic field can be developed, where the systematic error arises from the systematic deformation of the structure. Meanwhile, considering the random error, the electromechanical two-field coupling model can be derived as follows:

$$E(\theta, \phi) = \iint_A \frac{f_0(\xi - \Delta\xi(\delta(\beta)), \phi' - \Delta\phi'(\delta(\beta)))}{r_0}$$
$$\cdot \exp j[k\rho' \sin\theta \cos(\phi - \phi')]$$
$$\cdot \exp j[\varphi_f(\delta(\beta)) + \varphi_s(\delta(\beta)) + \varphi_r(\gamma)]\rho' d\rho' d\phi' \tag{3.17}$$

Figure 3.4 Dual reflector antenna.

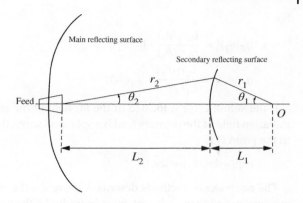

where $f_0(\xi - \Delta\xi(\delta(\beta)), \phi' - \Delta\phi'(\delta(\beta)))$ is the influence term of the feed pointing error on the amplitude of the aperture field caused by the structural displacement field of the reflector surface, $\varphi_f(\delta(\beta))$ is the influence term of the feed position error on the phase of the aperture field, $\varphi_s(\delta(\beta))$ is the influence term of the surface systematic deformation on the phase of the aperture field in the main reflecting surface, and $\varphi_r(\gamma)$ is the effect term of the random errors on the phase of the aperture field in the reflector panels.

3.2.5 Dual Reflector Antenna

The previous discussion is the electromechanical two-field coupling model of a feed-forward single reflector antenna. For the widely used Cassegrain dual reflector antennas, the electromechanical two-field coupling modeling process is similar with the single reflector antenna; the only difference is to determine the equivalent phase center of the dual reflector antenna. Therefore, in Figure 3.4, the feed and the secondary reflective surface are treated as equivalent to the radiation source at point O.

Assuming that the radiation pattern of the feed is described as $f(\theta_2)$ and the equivalent radiation pattern is $f(\theta_1)$, the following equations can be obtained from the power conservation condition:

$$|f_E(\theta_1)| = \frac{L_2}{L_1}\sqrt{\frac{\sin\theta_2 d\theta_2}{\sin\theta_1 d\theta_1}} \cdot |f_E(\theta_2)| \quad 0 \le \theta_1 \le \theta_m \tag{3.18}$$

$$|f_H(\theta_1)| = \frac{L_2}{L_1}\sqrt{\frac{\sin\theta_2 d\theta_2}{\sin\theta_1 d\theta_1}} \cdot |f_H(\theta_2)| \quad 0 \le \theta_1 \le \theta_m \tag{3.19}$$

Since

$$\begin{cases} ds = r_1^2 \sin\theta_1 d\theta_1 d\varphi = r_2^2 \sin\theta_2 d\theta_2 d\varphi \\ r_1 \sin\theta_1 = r_2 \sin\theta_2 \end{cases}$$

So

$$|f_E(\theta_1)| = \frac{L_2 r_1(\theta_1)}{L_1 r_2(\theta_2)} \cdot |f_E(\theta_2)| \tag{3.20}$$

$$|f_H(\theta_1)| = \frac{L_2 r_1(\theta_1)}{L_1 r_2(\theta_2)} \cdot |f_H(\theta_2)| \tag{3.21}$$

If the subreflector is located in the far field of the feed, and assuming that the radiation field of the primary feed is a spherical wave, the phase pattern equivalent to the point O is

$$\exp[-jkr_2(\theta_2) + jkr_1(\theta_1)] \tag{3.22}$$

The equivalence methods described above are the effectiveness for electromechanical coupling models with hyperbolic dual reflector or modified dual reflector.

3.2.6 Experiment

In order to confirm the correctness of the theoretical model of electromechanical field coupling, the experiments were conducted in an antenna performance test field located in Mei County, Shaanxi Province, China, and the details are described below.

3.2.6.1 Basic Parameters

Figure 3.5 shows a 3.7 m aperture, C/Ku dual-band, ring-focus, and Cassegrain-type reflector antenna. Its subreflector surface diameter is 0.44 m, the ratio of the focal length to diameter is 0.35, and the operating frequency is 12.5 GHz. The feed is a dielectric loaded horn. The aperture field distribution function for the antenna design is

$$f_{S'}(\vec{\rho}') = \begin{cases} 1 - 0.9 \exp[(\vec{\rho}')^2 - 1] & \vec{\rho}' \geq 0.5 \\ 1 - 0.85 \exp(0.13 - \vec{\rho}') & \vec{\rho}' < 0.5 \end{cases}$$

where $\vec{\rho}' \in [0, 1]$ is the normalized radius.

The main reflector surface of the antenna consists of 12 identical fan-shaped subpanels, and the backup frame consists of 12 amplitude beams, one ring beam, and a central body. The panel is connected to the radial beam by 13 bolts.

3.2.6.2 The Basic Idea of the Experiment

The purpose of this experiment is to verify the correctness of the theoretical model of the electromechanical field coupling model which will be proofed through practical measurements. To do this, the basic idea is as follows. First, the antenna reflector is artificially deformed and the deformation of the reflector is measured and recorded (500 typical points) with the corresponding far-field

Figure 3.5 A C/Ku-band 3.7 m aperture reflector antenna.

patterns (gain, first sidelobe levels, and 3 dB beamwidth). Second, the finite element model is established corresponding to the antenna structure, and the same structural displacement field is generated as the actual deformation, and then the corresponding electrical properties using the electromechanical coupling model equation (3.17) are derived, i.e. the far-field radiation pattern. Thirdly, the electrical performance calculated by the electromechanical field coupling theoretical model is compared with the measured electrical performance to obtain error I. Fourthly, for the reflecting surface antenna with surface error, the corresponding electrical performance is calculated by Feko and compared with the measured results to obtain error II, and then errors I and II are compared. Finally, the correctness and validity of the electromechanical field coupling theoretical model can be verified.

3.2.6.3 Working Conditions and Deformation

In order to deform the main reflecting surface, and considering that the amplitude distribution of the antenna's aperture field is in the form of "double peaks," a load is applied on the back of the panel to deform the panel. The star red dot on the panel in Figure 3.6 represents the position of the load application. Two normal displacements of 1.5 and 3 mm at the nodes of the reflective surfaces corresponding to the five bolts are created. The structural deformation corresponding to each of these conditions can be obtained by finite element analysis. The entire finite element model of the antenna structure consists of 7569 nodes and 1237 elements, which include 252 beam elements and 985 shell elements.

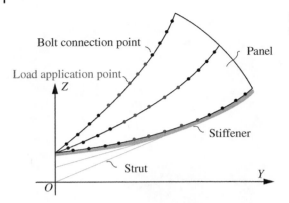

Figure 3.6 Schematic diagram of the load application position.

3.2.6.4 Measurement and Environment

Surface precision measurement of the reflective surface is performed by a laser range, and the electrical properties are obtained by a spectrometer. The temperature at the test site is 25 °C, and the wind speed is 4 m/s.

3.2.6.5 Calculated and Measured Results

Table 3.1 provides a comparison of the calculated and measured results in the original state (working condition 1) and two working conditions mentioned above. The calculation is carried out in two cases, one is obtained using the theoretical model of electromechanical coupling, and the other is obtained by Feko software. The results using the electromechanical coupling model and Feko software for each working condition separately are shown in the table, which are compared to the measured results. Using the electromechanical coupling model under the three working conditions, the maximum relative errors for the antenna gain, first sidelobe level, and beam width are obtained, which are separately 1.17%, 4.86%, and 2.63% relative to the measured results. It is clear that the calculated values of the electromechanical coupling model are much more closer to the measured results than that with the Feko software.

3.3 Planar Slotted Waveguide Array Antennas

Planar slotted waveguide array antennas are mainly used in airborne radars, such as the meteorological radar of the "Yun-20." The antenna consists of a radiation waveguide (front), a coupling waveguide (middle), and an excitation waveguide (back) (Figure 3.7), which is a thin-walled cavity structure. The manufacturing process is that, to begin with, each waveguide layer is machined separately by a high-precision machining. Then, the solder is placed between every two layers

Table 3.1 Analytical and measured results of the 3.7 m diameter reflector antenna under different working conditions

Cases	Performance	Gain (dB)	The relative error (%)	3 dB beam width (°)	The relative error (%)	First sidelobe on the left (dB)	The relative error (%)	First sidelobe on the right (dB)	The relative error (%)
Working condition No. 1	Measured	52.23		0.380		−14.54		−14.18	
	Coupling	52.84	1.17	0.390	2.63	−14.33	1.44	−14.33	1.06
	FEKO	52.92	1.32	0.390	2.63	−14.32	1.51	−14.34	1.13
Working condition No. 2	Measured	52.12		0.396		−15.24		−15.41	
	Coupling	52.71	1.13	0.399	0.76	−14.50	4.86	−14.67	4.80
	FEKO	52.77	1.25	0.400	1.01	−14.13	7.23	−14.17	8.05
Working condition No. 3	Measured	51.79		0.408		−14.81		−14.69	
	Coupling	51.90	0.21	0.403	1.22	−14.56	1.69	−14.07	4.22
	FEKO	52.13	0.66	0.401	1.72	−13.67	7.70	−13.26	9.73

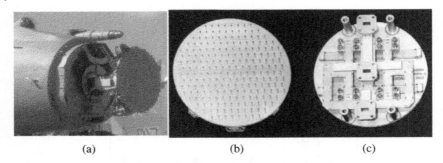

(a)   (b)   (c)

Figure 3.7 Planar slotted array antennas. (a) Physical view in service. (b) Radiation surface. (c) Back side. Source: (a) Valka.cz.

and bounded together. Followed by a certain clamping measure, the three layers are finally placed in a salt solution or vacuum welding. After that, the antenna is heated up to approximately 600°C and then gradually cooled down to room temperature in a gradient curve.

There are two main factors in this machining process that affect the final electrical performance: the machining accuracy of the single-layer waveguide itself, especially the machining errors in the numerous radiation slots of the radiation waveguide, and the errors caused by thermal processing. The former one is easy to ensure, because the machining accuracy of multiaxis machining centers can be very high and can meet the requirements now. The difficulty is the latter one, as hot machining will enlarge the previous error. How to make sure this process to guarantee the accuracy after hot machining is the most critical point, which will be discussed specifically later.

Back to the planar slotted array antenna in service, the planar slotted array antenna is usually a thin-walled, cavity, and multilayer structure. As these antennas are mainly used in airborne radar systems, their aperture is generally about 1 m, and the complicated working environment loads (vibration and shock, etc.) will easily cause structural deformation, which will lead to the position and direction errors of the radiation slot of the array as well as the changed mutual coupling effect, and eventually result in the failure of antenna electrical performance. Therefore, it is necessary to analyze the influence of the radiation slot position shift and pointing deflection on electronic performance, which are caused by the deformation duo to environment loads firstly. At the same time, considering the influence of the slot cavity deformation on the radiating slot voltage, an electromechanical two-field coupling theoretical model of the array structure displacement field and the antenna electromagnetic field is established [12, 13].

Figure 3.8 Geometric relationship of the planar slotted array antenna.

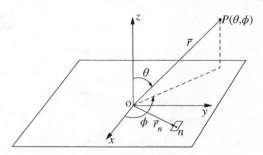

As shown in Figure 3.8, the radiation array of the antenna is located in plane xoy, the positive direction of the z-axis is the direction normal to the array, and the origin o is the geometric center of the array. The radiation array is a flat plate with a large number of radiation slots, through which the electromagnetic energy is radiated into the outer space. In order to clearly describe the coordinate relationship of the antenna, only the nth slot (radiation slot) is shown in Figure 3.8. The coordinate of the center of the gap is $\vec{r}_n = (x_n, y_n, 0)(n = 1, 2, \cdots, N)$, where N is the total number of gaps in the surface of the formation. According to antenna theory, in the direction of observation, the electromagnetic field of the planar slotted array antenna is expressed as

$$E(\theta, \phi) = \sum_{n=1}^{N} V_n \cdot f_n(l_n, w_n, \theta, \phi) \cdot e^{j\eta_n + j\, k\vec{r}_n \cdot \hat{r}} \tag{3.23}$$

where V_n and $\eta_n (n = 1, 2, \cdots, N)$ are the slot voltage amplitude and phase of the nth slot, respectively, $f_n(l_n, w_n, \theta, \phi) (n = 1, 2, \cdots, N)$ is the element pattern for the nth slot, l_n and w_n are the length and width of the nth slot, respectively, θ and ϕ are the directions of observation at point P in space, k is the propagation constant, and $\hat{r} = \vec{r}/r = (\sin\theta \cos\phi, \sin\theta \sin\phi, \cos\theta)$ is the unit vector of vector $\vec{r}$.

The specific functional form of $f_n(l_n, w_n, \theta, \phi)$ is

$$f_n(l_n, w_n, \theta, \phi) = jk(H_n(l_n, w_n, \theta, \phi) \sin\phi \hat{a}_\theta + H_n(l_n, w_n, \theta, \phi) \cos\phi \cos\theta \hat{a}_\phi) \tag{3.24}$$

$$H_n(l_n, w_n, \theta, \phi) = \frac{\frac{2\pi}{l_n} \cos\left(\frac{kl_n \sin\theta \cos\phi}{2}\right)}{\left(\frac{\pi}{l_n}\right)^2 - (k \sin\theta \cos\phi)^2} \cdot \frac{\sin(k \sin\theta \sin\phi w_n/2)}{k \sin\theta \sin\phi w_n/2} \tag{3.25}$$

where $\hat{a}_\theta, \hat{a}_\phi$ are the unit vectors in the direction of θ, ϕ in the spherical coordinate system. The seam voltage V_n will affect the amplitude and phase of point P, $f_n(l_n, w_n, \theta, \phi)$ only affects the amplitude of point P, and the position $\vec{r}_n$ of the radiation slot only affects the phase of point P.

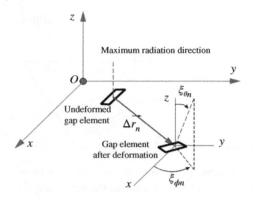

Figure 3.9 Geometric relationships after antenna deformation.

3.3.1 Effect of Position Error of the Radiation Slot

Equation (3.36) is the far-field pattern derived by assuming that the planar slotted array antenna is an ideal flat plate. However, in practical engineering, the antenna will be deformed under external loads such as vibration and impact, i.e. the radiation slots will not be in the same plane. When a planar slotted array antenna is subjected to an environmental load, the array radiation slot position deflection and pointing deflection can be obtained through structural analysis. As shown in Figure 3.9, assuming that the nth radiating slot moves by $\Delta \vec{r}_n$ from its original position, the normal of the slot will be rotated at $\xi_{\theta n}, \xi_{\phi n}$.

In the ideal case, the spatial phase information for the nth radiation slot is obtained from Eq. (3.36) as

$$\varphi_n = k\vec{r}_n \cdot \hat{r} \tag{3.26}$$

When subjected to a load, the nth radiation slot will move by $\Delta \vec{r}_n$ from its original position, and the spatial phase becomes

$$\varphi'_n(\delta(\beta)) = k(\vec{r}_n + \Delta\vec{r}_n(\delta(\beta))) \cdot \hat{r} = \varphi_n + \Delta\varphi_n(\delta(\beta)) \tag{3.27}$$

If the random errors γ in the manufacturing and assembly process of the antenna are taken into account, the new spatial phase φ'_n of the nth radiating slot can be rewritten as

$$\varphi'_n(\delta(\beta), \gamma) = \varphi_n + \Delta\varphi_n(\delta(\beta), \gamma) \tag{3.28}$$

3.3.2 Effect of Radiation Slot Pointing Deflection

If the nth radial slot is rotated by $\xi_{\theta n}$ and $\xi_{\phi n}$ along the coordinate axis, then the new radial slot radiation pattern for the nth radial slot becomes

$$f_n(l_n, w_n, \theta, \phi, \delta(\beta)) = f_n(l_n, w_n, \theta - \xi_{\theta n}(\delta(\beta)), \phi - \xi_{\phi n}(\delta(\beta))) \tag{3.29}$$

The new pattern of the nth radiating slot can be rewritten, when the random errors γ in the manufacturing and assembly process of the antenna are taken into account, as the following form:

$$f_n(l_n, w_n, \theta, \phi, \delta(\beta), \gamma) = f_n(l_n, w_n, \theta - \xi_{\theta n}(\delta(\beta), \gamma), \phi - \xi_{\phi n}(\delta(\beta), \gamma)) \quad (3.30)$$

3.3.3 Effect of Seam Cavity Deformation on Radiation Seam Voltage

Under the influence of external loads, the planar slotted array antenna is deformed, and the internal slot cavity structure consisting of slot and cavity will also be deformed at the same time, which will eventually bring about changes in the radiation slot voltage. Firstly, the deformation of the antenna slot cavity brings about changes in the electromagnetic field guide path, causing changes in the radiation slot voltage. Then, the deformation of the array surface will cause the change in mutual coupling outside the radiation slot, which will affect the change of the radiation slot voltage meanwhile. Therefore, the amplitude and phase of the radiation slot voltage under the influence of the deformation of the slot cavity can be described as

$$V_n'(\delta(\beta)) = V_n + \Delta V_n(\delta(\beta)) \quad (n = 1, 2, \cdots, N) \quad (3.31)$$

$$\eta_n'(\delta(\beta)) = \eta_n + \Delta \eta_n(\delta(\beta)) \quad (n = 1, 2, \cdots, N) \quad (3.32)$$

In addition, taking the random error γ into account during the manufacturing and assembly of the seam cavity, the new radiation seam voltage amplitude and phase can be rewritten as

$$V_n'(\delta(\beta), \gamma) = V_n + \Delta V_n(\delta(\beta), \gamma) \quad (n = 1, 2, \cdots, N) \quad (3.33)$$

$$\eta_n'(\delta(\beta)) = \eta_n + \Delta \eta_n(\delta(\beta), \gamma) \quad (n = 1, 2, \cdots, N) \quad (3.34)$$

3.3.4 Two-field Electromechanical Coupling Model

Combining the above factors, a theoretical model of the field coupling between the electromagnetic field and the displacement field of a planar slotted array antenna can be derived as

$$E(\theta, \phi) = \sum_{n=1}^{N} A_n \cdot f_n(l_n, w_n, \theta, \phi, \delta(\beta), \gamma) \cdot V_n'(\delta(\beta), \gamma) e^{j\eta_n'(\delta(\beta), \gamma)} \cdot e^{j\varphi_n(\delta(\beta), \gamma)}$$

$$(3.35)$$

where $A_n = e^{jk\vec{r}_n \cdot \hat{r}_0}$, $\delta(\beta)$ is the structural displacement field corresponding to the antenna deformation, β is the structural design variable, γ is the random error caused during manufacturing and assembly, etc., $f_n(l_n, w_n, \theta, \phi, \delta(\beta), \gamma)$, $V_n'(\delta(\beta), \gamma)$, and $\eta_n'(\delta(\beta), \gamma)$ are the element pattern, slot voltage amplitude, and slot voltage phase change terms caused by the antenna deformation and random error, respectively, and $\varphi_n(\delta(\beta), \gamma)$ is the spatial phase of the mouth surface.

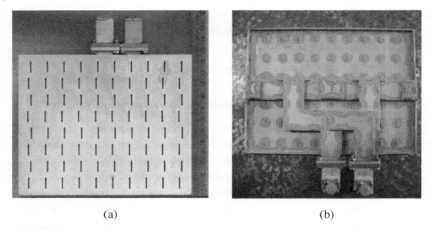

(a)   (b)

Figure 3.10   Structure of a planar slotted array antenna. (a) Front view. (b) Back view.

3.3.5   Experiment

In order to confirm the correctness and validity of the theoretical model of field coupling (3.35), an example of a 15 GHz planar slotted array antenna shown in Figure 3.10 is given as follows.

3.3.5.1   Basic Parameters

The bomb-loaded planar slotted array antenna consists of a radiation layer, a coupling layer, an excitation layer, and a sum and difference network and contains two subarrays, each consisting of four radiating waveguides, one coupling waveguide, and one excitation waveguide, where each radiating waveguide has eight radiation slots. The coupling waveguide has five coupling slots, and the excitation waveguide has one excitation slot. The 8×10 array of the radiant panels is formed by radiation slots, and its array aperture size is 150×126 mm. Its thickness is 1 mm. The wall thickness between the layers of the waveguide is 1 mm, and the thickness of the outer waveguide edge of the antenna is 2 mm. The specific dimensional information of each layer of waveguide cavity is shown in Table 3.2.

The antenna is used as a bomb-guided head antenna, operating in Ku-band with a center frequency of 15 GHz and a bandwidth of 400 MHz, requiring an antenna gain of not less than 25 dB, a maximum sub-sidelobe level of not more than −17 dB, and a voltage standing wave ratio of less than 1.5.

According to the characteristics of the thin-walled, cavity structure of the antenna, the finite element model shown in Figure 3.11 is built, i.e. 36 127 triangular shell units are used, with a total of 18 137 nodes. Each node consists of 6 degrees of freedom.

Table 3.2 Structural parameters of each waveguide cavity.

Size	Radiation waveguide	Coupling waveguide	Incentive waveguide	Sum and difference network
Width (mm)	13.5	13.81	14.78	13.78
Height (mm)	4	4	4	4

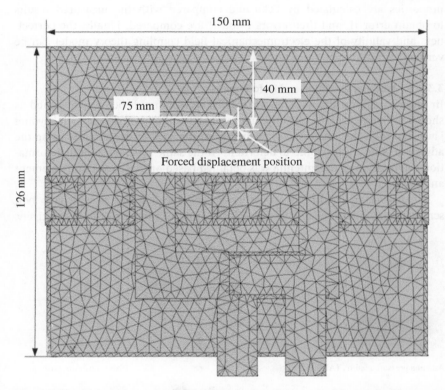

Figure 3.11 Finite element model and the location of the applied forced displacement.

For the purpose of artificially generating errors in the array, the load is applied at the location where the coordinate values are marked as shown in Figure 3.11.

3.3.5.2 Basic Idea

The experiment consists of the following steps. Firstly, an artificial error is generated in the antenna array, and the error value (80 points) is measured and recorded together with the corresponding far-field radiation pattern (gain, maximum sub-band level in azimuth and pitch plane, and 3 dB wavefront width). Secondly, a finite element model corresponding to the antenna structure is built, and the same error as the actual surface is generated, and the corresponding electrical performance, i.e. the far-field radiation pattern, is derived using the electromechanical field coupling theory model (3.35). Thirdly, the electrical properties calculated by the electromechanical field coupling theory model are compared with the measured electrical properties to obtain error I. Fourthly, for the planar slotted array antenna with profile error, the corresponding electrical properties are calculated by Feko and compared with the measured results to obtain error II, and then errors I and II are compared. Finally, the correctness and validity of the electromechanical field coupling theory model can be verified.

3.3.5.3 Working Condition and Deformation

In order to deform the antenna artificially, a clamping device was designed as shown in Figure 3.12, i.e. the antenna was fixed at the four corners with bayonet jaws, and then the antenna was pushed from the back of the antenna using the adjusting bolts on the clamp to deform it. At the same time, the overall deformation information of the antenna was obtained based on structural finite element analysis.

Three working conditions are designed, and the first one is with zero bolt screw-in, and the second and third are with 1.0 and 1.5 mm screw-in, respectively.

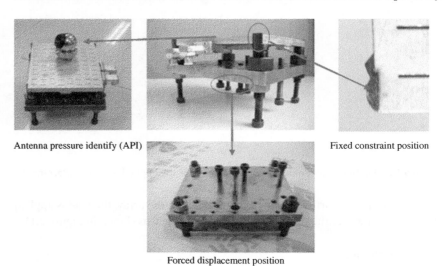

Antenna pressure identify (API)　　　　　　　　　　　　Fixed constraint position

Forced displacement position

Figure 3.12　The mounting device to force the antenna deformed.

Table 3.3 Analysis and experimental results of a planar slotted array antenna with a frequency of 15 GHz under different working conditions.

Cases	Performance	Gain (dB)	The relative error (%)	Azimuth plane				Elevation plane			
				Maximum sidelobe (dB)	The relative error (%)	3 dB Beam width (°)	The relative error (%)	Maximum sidelobe (dB)	The relative error (%)	3 dB Beam width (°)	The relative error (%)
Working condition No. 1	Measured	26.20		−18.48		9.25		−22.78		7.79	
	Coupling	26.46	0.99	−17.80	3.68	9.20	0.54	−22.73	0.22	8.02	2.95
	FEKO	26.54	1.30	−17.74	4.16	9.18	0.76	−22.70	0.35	8.04	3.21
Working condition No. 2	Measured	26.14		−16.95		9.23		−22.20		8.10	
	Coupling	26.45	1.19	−16.82	0.77	9.48	2.71	−21.69	2.30	8.21	1.36
	FEKO	26.48	1.30	−16.34	3.60	9.55	3.47	−21.14	4.77	8.29	2.35
Working condition No. 3	Measured	26.06		−17.96		9.12		−22.42		8.05	
	Coupling	26.44	1.46	−16.81	6.40	9.31	2.08	−21.83	2.63	8.22	2.11
	FEKO	26.47	1.57	−16.32	9.13	9.38	2.85	−21.75	3.00	8.46	5.09

3.3.5.4 Testing and Environment

The mechanical structure and electrical performance measurements were carried out indoors. The array deformation was measured by laser ranges, and the electrical properties were measured in a microwave darkroom at the State Key Laboratory of Antennas and Microwaves, Xi'an University, China.

3.3.5.5 Calculated and Measured Results

Table 3.2 lists a comparison of the calculated and measured results for the three working conditions. The calculations are carried out in two cases, one is obtained using the theoretical model of electromechanical coupling, and the other is obtained using the Feko software. A comparison of the results obtained by electromechanical coupling model and Feko software with the measured results is presented for each working condition. The maximum relative errors of the electromechanical coupling model in relation to the measured results for the three operating conditions are 1.46% for gain, 6.40% for maximum sidelobe level, and 2.95% for 3 dB beam width. The maximum relative errors of the Feko software in relation to the measured results for the three working conditions were 1.57% for gain, 9.13% for maximum sidelobe level, and 5.09% for 3 dB beam width. Clearly, the calculated values obtained by the electromechanical coupling model are closer to the measured results than that with the Feko software; thus, the correctness of the electromechanical coupling model is verified (Table 3.3).

3.4 Active Phased Array Antennas

Active phased array antenna (APAA) is widely used in radar, communication, detection, and other systems in land, sea, air, and sky fields. Figure 3.17a shows the active phased array antenna of the mobile land-based anti-aircraft and anti-missile radar, which works in X-band and has high requirements on array planarity, pointing accuracy, multiple targets, etc. Since the radar beam is required to achieve all-round scanning, it adopts mechanical sweep and electric sweep. The combination of mechanical and electrical scanning is required. In terms of the accuracy of the radar array, there are two ways that need to be seriously considered, one is the error introduced in the manufacturing and assembly process from the radiation component to the module, the module to the subarray, and the subarray to the large array in the upper right corner, and the other is the error introduced in the assembly process from the wheel track to the seat frame, the back frame, and the large array in the lower left corner.

At the same time, there are a large number of heat-generating devices in the array. There are some transmit and receive T/R components among them, which are particularly sensitive to temperature. The uneven temperature distribution of

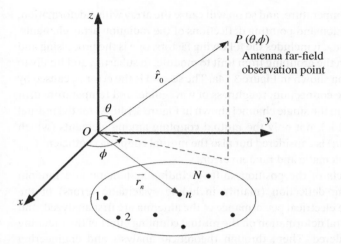

Figure 3.13 Array spatial coordinate relationships of APAA.

the array will affect the phase control accuracy of the antenna array. The complex operating environment load (vibration, shock, etc.) and temperature distribution will cause structural deformation, which will lead to changes in the radiation pattern of the radiation array elements and the mutual coupling effect between them, eventually leading to the failure of the antenna electrical performance to meet the requirements or even to be unachievable [14, 15]. The following section discusses the establishment of the theoretical model of field coupling among the APAA electromagnetic field, displacement field, and the temperature field.

As shown in Figure 3.13, there is an APAA, where the total number of radiation array elements is N, the nth radiation array element excitation current is $I_n \exp(j\varphi_{I_n})\hat{\tau}_n$, $\hat{\tau}_n$ is the unit polarization unit vector, and I_n φ_{I_n} are the amplitude and phase, respectively. If the nth array radiation pattern element in the array is $f_n(\theta, \phi)$, the position vector is $\vec{r}_n = x_n\hat{i} + y_n\hat{j} + z_n\hat{k}$, the far observation direction is $P(\theta, \phi)$, and the electrical field of the active phased array antenna can be expressed as

$$E(\theta, \phi) = \sum_{n=1}^{N} A_n \cdot f_n(\theta, \phi) \cdot I_n e^{j\varphi_{I_n}} \qquad (3.36)$$

where A_n is the spatial phase factor of the array element n, and the unit vector of the observation direction $P(\theta, \phi)$ is $\hat{r}_0 = (\sin\theta\cos\phi, \sin\theta\sin\phi, \cos\theta)^T$.

Equation (3.36) is the far-field radiation pattern when APAA is assumed to be an ideal array, and the mutual coupling between the elements is not taken into account. In fact, there are not only systematic and random errors but also mutual coupling between the elements. In terms of systematic errors, the loads of wind, snow, self-weight, temperature, and so on will cause the array with δ deformation, resulting in the position and pointing deflections of the radiation array elements. For the random error γ, it includes the following factors: one is the processing and assembly errors γ_1 in the grouping from unit to module, to subarray, and finally to large array of antenna shown in Figure 3.14a. The second is the error γ_2 caused by discontinuous flange connection, roughness of waveguide, and temperature drift of T/R performance in the single channel shown in Figure 3.14b. As for the mutual coupling (Figure 3.14c), not only the mutual coupling among elements (which may be set to n and m) is considered but also the mutual coupling coefficient C_{nm} is related to both systematic and random errors.

For this, the effects of the position shift (including systematic and random errors) and pointing deflection (mainly including systematic errors) of the radiating unit on the electrical performance of the antenna are first analyzed, and the effects of structural deformation on the mutual coupling effect of the radiating unit are also considered. Then, through theoretical analysis and engineering tests, the effects of temperature variations and the excitation current amplitude and phase of the element mutual coupling T/R components are investigated. Finally, the APAA electromechanical–thermal three-field coupling model, which integrates the displacement, electromagnetic, and temperature fields of the array, is established [16–19].

3.4.1 Effect of Change of Position and Attitude of the Radiation Unit

Without loss of generality, it may be useful to set the position offset of the nth element in Figure 3.15 as $\Delta \vec{r}_n$, and the pointing deflections (i.e. the change in maximum radiation direction) are $\xi_{\theta n}$ and $\xi_{\phi n}$.

In the ideal case, the spatial phase factor for the nth ($n = 1, 2, \ldots, N$) radiating unit is obtained from Eq. (3.35) as

$$A_n = \exp(jk\vec{r}_n \cdot \hat{r}_0) \tag{3.37}$$

If its position deviates from the design value to $\Delta \vec{r}_n$, the spatial phase factor becomes

$$A'_n = \exp\{jk[\vec{r}_n + \Delta \vec{r}_n(\delta(\beta, T))] \cdot \hat{r}_0\}$$
$$= \exp(jk\vec{r}_n \cdot \hat{r}_0) \cdot \exp(jk\Delta \vec{r}_n(\delta(\beta, T)) \cdot \hat{r}_0) = A_n \cdot \exp\left(j\varphi'_n(\delta(\beta, T))\right) \tag{3.38}$$

where $\delta(\beta, T)$ is the position error of the radiating unit caused by the environmental load and temperature (i.e. systematic error), $\beta = (\beta_1, \beta_2, \cdots, \beta_{Nd})$ is the structural design variable, and T is the temperature distribution of the antenna array.

Figure 3.14 Active phased array antennas with construction (a), single excitation (b), and mutual coupling (c).

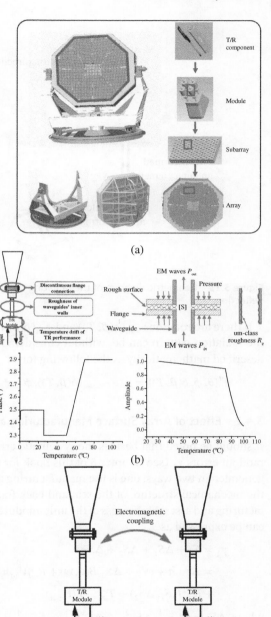

(a)

(b)

(c)

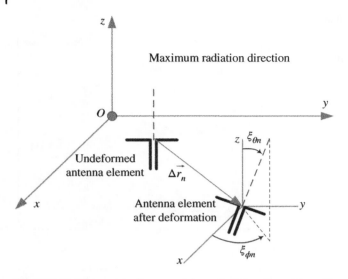

Figure 3.15 Geometric diagram of the radiating unit position offset and pointing deflection.

If there are rotations $\xi_{\theta n}(\delta(\beta, T))$ and $\xi_{\phi n}(\delta(\beta, T))$ of the nth radiating element, the radiation pattern can be, without consideration of mutual among elements, described mathematically as the following form:

$$f_n'(\theta, \phi, \delta(\beta, T)) = f_n(\theta - \xi_{\theta n}(\delta(\beta, T)), \phi - \xi_{\phi n}(\delta(\beta, T))) \tag{3.39}$$

3.4.2 Effect of Array Surface Manufacturing and Assembly Errors

It should be noted that besides the systematic errors mentioned above, there are random errors γ_1 (see Chapter 5 of this book for a detailed derivation), which is generated in two ways, one is the manufacturing accuracy and assembly errors of the mechanical structure of the seat and back frame, and the other is the manufacturing and assembly process of the unit–module–subarray–array surface, which can be expressed as

$$\gamma_1 = S_0 + \Delta S_1 + \Delta S_2 + \Delta S_3$$
$$= \Delta\varsigma \cdot \vec{n} + (\tilde{V}^e + \Delta\vec{S}_0)B(u, w) + R_{hd}V_{hd}h_{hd}$$
$$+ K^{-1}f(S_1, \Delta\vec{S}_2) + T_{m,n} \cdot \Gamma(P_{m,n}) \tag{3.40}$$

where $\Delta S_1 = [\Delta S_1^{1,1}, \Delta S_1^{1,2}, \cdots \Delta S_1^{M,N}]$, $\Delta S_1^{m,n} = [T_{m,n}, P_{m,n}]^T$ ($m = 1, 2, ..., M$; $n = 1, 2, ..., N$), $\Delta S_2 = p_{hd}(u, w) = p_{id}(u, w) + C_{hd}(u, w) \cdot h_{fd}(u, w)$, $\Delta S_3 = \Delta\varsigma \cdot \vec{n}$, $\tilde{V}^e$ and $B(u, w)$ are the control vertices and base function of the dual cubic B-sample surface, $T_{m,n}$ is the localized surface rotation of the subarray, $P_{m,n}$ is the transformation vector relative to the position of each rotation, $p_{id}(u, w)$ is the integer dimensional surface component, $C_{hd}(u, w)[w1]$ is the mixed dimensional surface correlation coefficient, $h_{fd}[w1]$ is the fractional dimensional height, ΔS_3 is the projection of the random error $\Delta\varsigma$ in the normal direction $\bar{n}$ of each radial array element due to the accumulation of base support errors, K is the stiffness matrix, and f is the assembly force.

Thus, Eqs. (3.38) and (3.39) can be further written as

$$A'' = A_n \cdot \exp\left(j\varphi_n''(\delta(\beta, T), \gamma_1)\right) \tag{3.41}$$

$$f_n''(\theta, \phi, \delta(\beta, T), \gamma_1) = f_n(\theta - \xi_{\theta n}(\delta(\beta, T), \gamma_1), \phi - \xi_{\phi n}(\delta(\beta, T), \gamma_1)) \tag{3.42}$$

Thus, Eq. (3.35) can be similarly further written as

$$E(\theta, \phi) = \sum_{n=1}^{N} A_n \cdot \exp\left(j\varphi_n'(\delta(\beta, T), \gamma_1)\right)$$

$$f_n(\theta - \xi_{\theta n}(\delta(\beta, T), \gamma_1), \phi - \xi_{\phi n}(\delta(\beta, T), \gamma_1)) \cdot I_n e^{j\varphi_{I_n}} \tag{3.43}$$

3.4.3 Effect of Radiation Array Element Manufacturing and Assembly Errors

As mentioned previously, for a single radiation channel, the structural factors affecting the electrical performance are mainly flange connection discontinuities, waveguide inner wall roughness, and T/R component temperature drift (Figure 3.14b), which could be described separately below.

3.4.3.1 Waveguide Flange Connection Discontinuity

The transmission characteristics of waveguides are easily affected when flange connections are encountered during the transmission of electromagnetic waves. Considering the inevitable error, i.e. roughness of the interconnected flange faces, the manufacturing error, results in three cases between the faces as shown in Figure 3.16a, namely, direct metal-to-metal contact MM, noncontact air, and oxide insulation MIM. In order to derive the equivalent impedance, the impedance model shown in Figure 3.16b is developed, where R_{MM} and R_{MIM} are the resistances at the corresponding locations. The physical quantities per unit depth of the flange surface contact structure along the electromagnetic wave transmission direction can be derived. The inductance is

$$L = \mu_0 \cdot (d/l_w) \tag{3.44}$$

The capacitance of the oxide insulation layer is

$$C_c = \varepsilon l_w A_{MIM}/t \tag{3.45}$$

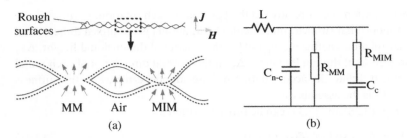

Figure 3.16 Equivalent impedance model. (a) Contact structural model. (b) Equivalent circuit of contact structure.

The air capacitance of the uncontacted part is

$$C_{n-c} = \varepsilon_0 \cdot (l_w/d) \tag{3.46}$$

in which d is the average spacing between the two sides of the flange, which will be changed as the bolt force F. The unit depth mentioned above corresponds to $d = 1$. Parameter l_w is the side length of the rectangular cross section of the waveguide, μ_0 is the air permeability, ε and ε_0 are the actual and air dielectric constants, respectively, t and A_{MIM} are the thickness and area of the oxide insulation layer, respectively.

Thus, the equivalent impedance is

$$Z_c = \sqrt{\dfrac{j\omega L}{j\omega C_{n-c} + \dfrac{1}{R_{MM}} + \dfrac{1}{R_{MIM} + (1/j\omega C_c)}}} \tag{3.47}$$

3.4.3.2 Influence of Waveguide Inner Wall Roughness

In general, the roughness of the inner wall of a metal waveguide is closely related to the processing. If the waveguide is mechanically drawn, the roughness is on the order of microns, so the root mean square (RMS) of the rough surface height is roughly equivalent to the skin depth in the GHz band, which causes transmission power loss and phase delay. How can this effect be described quantitatively? The equivalent impedance of a rough surface can be obtained by considering the rough metal surface as a superposition of multiple smooth layers with varying conductivity [20] and using a layered medium model to analyze the distribution of electromagnetic fields within the rough surface (Figure 3.17).

The final effect of the inner wall roughness on the transmission characteristics of the waveguide is obtained (Figures 3.18 and 3.19). Figure 3.18 describes the effect of the root mean square error of the inner wall roughness (1, 3, and 5 μm corresponding to the black (square), red (round spot), and blue (triangle) curves in the figure) on the S_{21} value (left panel) and the phase lag (right panel), respectively,

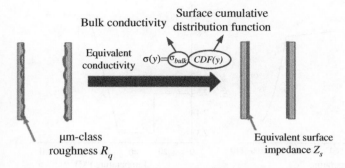

Figure 3.17 Equivalent diagram of waveguide inner wall roughness.

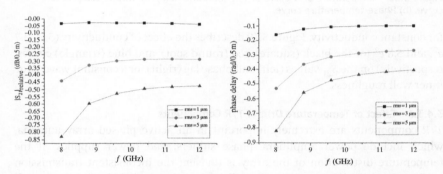

Figure 3.18 Effect of different inner wall roughness on the transmission performance with the same conductivity.

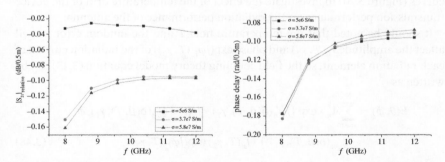

Figure 3.19 Effect of different metal conductivities on the transmission performance with the same roughness.

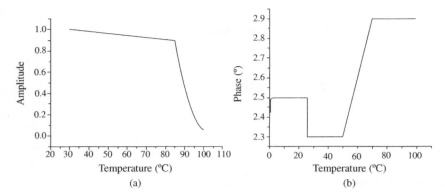

Figure 3.20 Temperature drift curve of T/R component. (a) Amplitude-temperature curve. (b) Phase-temperature curve.

for constant conductivity. Figure 3.19 describes the effect of conductivity ($5\,e^6$, $3.7\,e^7$, and $5.8\,e^7$ for the black (square), red (round spot), and blue (triangle) curves, respectively) on the S_{21} values (left) and phase lag (right) for a constant waveguide inner wall roughness.

3.4.3.3 Effect of Temperature Drift of T/R Components

T/R components are extremely important in an active phased array antenna, which include power amplifiers, phase shifters, and power supplies. If the temperature distribution of the array is uneven, the inconsistent transmission performance of many components will lead to the degradation of the overall radiation performance of the antenna. Here, the amplitude, phase, and temperature dependence of the output excitation of the T/R components are provided as curves (Figure 3.20) to investigate the effect of the temperature drift of the device transmission performance on the radiation performance of the antenna.

It should be noted that the temperature field T and the random error γ_2 will affect the amplitude $I_n(T, \gamma_2)$ and phase $\exp(j\varphi_{I_n}(T, \gamma_2))$ of the radiation current of each radiation element, so the field coupling theory model equation (3.43) can be written as

$$
E(\theta, \phi) = \sum_{n=1}^{N} A_n \cdot \exp\left(j\varphi_n'(\delta(\beta, T), \gamma_1) \cdot f_n(\theta - \xi_{\theta n}(\delta(\beta, T), \gamma_1), \phi\right.
$$
$$
\left. -\xi_{\phi n}(\delta(\beta, T), \gamma_1))\right) \cdot I_n(T, \gamma_2) \exp(j\varphi_{I_n}(T, \gamma_2)) \tag{3.48}
$$

3.4.4 Effect of Mutual Coupling of Radiation Elements on the Radiation Performance of Antennas

Radiation elements located in an array exhibit very different electromagnetic properties in the array environment and in isolation due to the mutual coupling

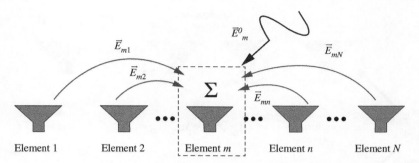

Figure 3.21 Electromagnetic environment of elements in the array.

effect among them. The mechanism, pathways, and quantitative calculation of the mutual coupling are important factors in the design of active phased array antennas.

Taking the array element m as an example, in addition to the incident electric field outside the array, it is also subjected to the electromagnetic scattering from the other elements (Figure 3.21). The sum of the above two parts is the total electric field, i.e.

$$\vec{E}_{inc}^{m} = \vec{E}_{m}^{0} + \sum_{\substack{n=1 \\ n \neq m}}^{N} \vec{E}_{mn} \tag{3.49}$$

There are various methods to calculate the mutual coupling effect, for instance the dipole method. It is clear from the eigenmode theory that when the operating frequency and structure are determined, the eigenmode of each order of the array element is determined, and the electrical properties of the array element can be expressed as a linear combination of modes from all sectors (Figure 3.22). The procedure is same as the displacement of a mechanical structure, which can be expressed as a linear combination of structural eigenvectors. Therefore, the determination of the corresponding eigenmode excitation coefficients of each order becomes critical [21].

The mutual coupling between different elements can be started from the perspective of mode coupling (Figure 3.23). If the electromagnetic response between different modes is used to measure the strength of mode coupling, the equilibrium equation for mode coupling of array elements in the array environment can be established as

$$\mathbf{V} = \mathbf{V}_0 + \mathbf{C}\Lambda\mathbf{V} \tag{3.50}$$

where $\mathbf{V}_0$ is the initial array of mode excitation coefficients for the full array element, $\mathbf{V}$ is the array of mode excitation coefficients in the full array element, Λ is the diagonal matrix associated with the eigenvalues of the array element, and

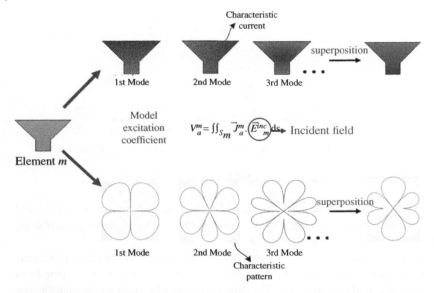

Figure 3.22 Modal decomposition of electromagnetic properties of array elements.

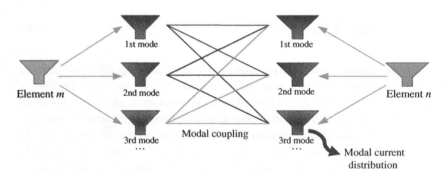

Figure 3.23 Characteristic modal coupling.

C is the mode coupling matrix, the element of which is subjected to

$$C_{ab}^{mn} = -j\omega\mu \int_{S_m}\int_{S_n} \chi_{ab}^{mn} g(r_{nm}(\gamma_1, \delta, T), \delta(\beta, T))dsds \tag{3.51}$$

where j is an imaginary unit, ω is the angular frequency, μ is the magnetic permeability, χ_{ab}^{mn} is the covariance associated with the mode current, and $g(r_{nm}, \delta(\beta, T))$ is the scalar Green's function, which is related to the array element relative position vector r_{nm}, the structural displacement field δ, and r_{nm} is related to γ_1, δ, and T. The coefficient of mutual coupling of the array elements can be obtained

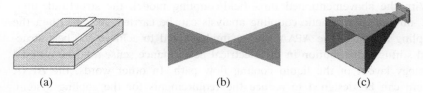

Figure 3.24 Three typical radiation array element forms. (a) Patch antennas. (b) Butterfly antennas. (c) Horn antennas.

from Eq. (3.51), and

$$
C_{nm}(\gamma_1, \delta, T) = \begin{cases} 1 & m = n \\ \left(\displaystyle\sum_{b=1}^{M} \sum_{a=1}^{M} \frac{I_b^c I_a^c}{1 + \eta_b^2} \frac{C_{ba}^{nm}}{1 + j\eta_a} \right) \Big/ \left(\displaystyle\sum_{b=1}^{M} \frac{(I_b^c)^2}{1 + \eta_b^2} \right) & m \neq n \end{cases} \tag{3.52}
$$

where η_a is the eigenvalue, and I_a^c is the array port mode current intensity.

As mentioned above, the total number of radiation arrays is N, and the total number of major current eigenmodes for each array is M. To derive the current patterns, the arrays can be dissected as a grid, and it is useful to set the total number of grid nodes to N_e, and in general, $M \ll N_e$.

As the current mode is only related to the structure of the radiating array element itself and not to the port excitation or the environment, the mode of an engineering structure is only related to the structure itself. For the different radiation array elements (Figure 3.24), only their main current modes need to be calculated, and the radiation characteristics of the whole array are calculated without further finite element mesh discretization of the array elements, so the computational effort $O((NM)^3)$ is much less than that of the full-wave method $O((NN_e)^3)$, which is due to the significantly lower dimensionality.

3.4.5 Theoretical Model of Electromagnetic–Displacement–Temperature Fields Coupling

In summary, if all errors with mutual coupling among elements are considered, the field coupling theoretical model of the electromagnetic–displacement–temperature fields of APAA in Eq. (3.46) can be transformed into the following form:

$$
\begin{aligned}
E(\theta, \phi) = \sum_{n=1}^{N} \Bigg[& \sum_{m=1}^{N} C_{nm}(\delta(\beta, T), T, \gamma_1) I_m(T, \gamma_2) \exp\left(j\varphi_{I_m}(T, \gamma_2)\right) \Bigg] \\
& \cdot f_n(\theta - \xi_{\theta n}(\delta(\beta, T), \gamma_1), \phi - \xi_{\phi n}(\delta(\beta, T), \gamma_1)) \cdot A_n \\
& \cdot \exp\left(j\varphi_n'(\delta(\beta, T), \gamma_1)\right)
\end{aligned} \tag{3.53}
$$

With the abovementioned three-field-coupling model, the structural, thermal, and electromagnetic coupling analysis can be carried out, and then the coupling design of the APAA can be implemented to achieve the best structural stiffness distribution in the electrical performance sense and the optimal topology layout of the liquid cooling flow path. In other words, the APAA system can be designed to reduce the requirements for the cooling system, structural machining accuracy, welding accuracy, and assembly accuracy under the same electrical performance index requirements. Under the same cooling system parameters and structural accuracy requirements, the cooling efficiency is improved, the structural weight and loop control requirements are reduced, and the overall performance of the active phased array antenna can be improved.

3.4.6 Experiment

In order to verify the correctness of the field coupling theory model (3.53), an experimental test platform was developed. Its basic components are shown in Figure 3.25, which mainly consists of the experimental antenna, control cabinet, test platform, cooling unit, external power supply, etc. [22].

3.4.6.1 Basic Parameters

The experimental antenna is an X-band active phased array antenna (center frequency 10 GHz and bandwidth 4 GHz), as shown in Figure 3.26 ((a) is the overall diagram, and (b) is the back-frame support and drive unit diagram). The antenna has a total number of 256 radiating array elements with a physical size of 2880 mm × 1728 mm × 1000 mm.

The experimental antenna consists of radiation array elements, a panel, a frame, active subarray modules (with T/R assembly), adjustment mechanisms, and other components, as shown in Table 3.4.

The active phased array subarray module is the core of the entire experimental platform hardware, consisting of a structural frame (including liquid-cooling runners), T/R components, and functional modules such as power supply, signal processing, and power divider. Its liquid-cooling runners are interconnected with the external cooling unit, and the components are cooled through the liquid-cooling runners inside the frame. The structural frame is the basis for the installation of equipment (including components) inside the subarray, and the frame and liquid-cooling runners adopt an integrated design, and the flow paths are formed inside the frame (Figure 3.27).

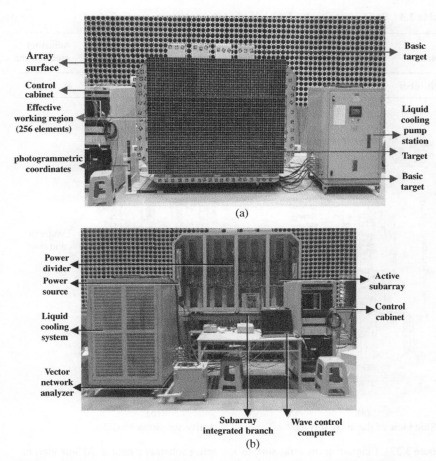

Figure 3.25 Experimental test platform systems. (a) From the front view, (b) from the back view.

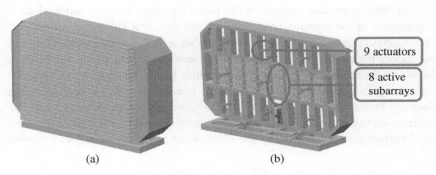

Figure 3.26 CAD model of an X-band active phased array antenna. (a) Overall structure. (b) The back-frame support and drive unit.

Table 3.4 Basic components of the experimental antenna.

Name	Array elements	Cables	T/R components	Active subarray	Array frame	Cooling units	Adjusting mechanisms
Number	256	256	256	8	1	1	9

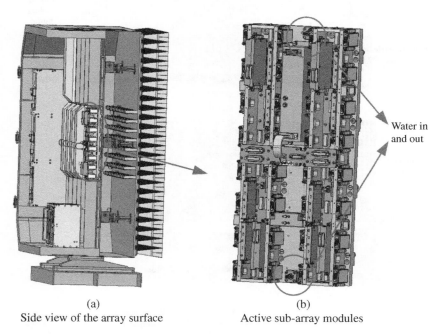

Water in
and out

(a)
Side view of the array surface

(b)
Active sub-array modules

Figure 3.27 Diagram of the array surface and active subarray modules. (a) Side view of the array surface. (b) Active subarray modules.

3.4.6.2 Basic Ideas

The experiment process consists of the following steps: first, an artificial error is generated in the antenna array surface, and the error value (measured with photography) is measured and recorded together with the corresponding far-field radiation pattern (gain, secondary sidelobe level, and 3 dB beam width, etc.). Second, a finite element model corresponding to the antenna array structure with the same error as the actual surface is created, and the corresponding electrical performance, i.e. the far-field radiation pattern, is obtained using the electromechanical–thermal field coupling theory model (3.53). Third, the electrical performance calculated by the electromechanical–thermal field coupling theory model is compared with the measured electrical performance to verify the correctness of the electromechanical–thermal field coupling theory model.

3.4.6.3 Working Conditions and Array Surface Errors

To generate the array surface error artificially, a worm gear as shown in Figure 3.26b is designed and fabricated. The error is generated by driving the antenna from the back using a worm gear on the frame. At the same time, the corresponding error information of the antenna is obtained based on the structural finite element analysis.

The experiment is carried out in three working conditions, i.e. changing the middle one of the three adjusting bolts on the left and the right to produce the required deformation of the array surface. The first working condition is that these two bolts were screwed in 1.0 mm, and the second and the third working conditions are screwed in 2.0 and 3.0 mm, which are 1/30, 1/15, and 1/10 of the wavelength, respectively.

3.4.6.4 Measurement and Environment

The mechanical structure and electrical performance measurements were carried out indoors (Figures 3.28 and 3.29). The array surface deformation was measured

24 × 32 scale experimental platform

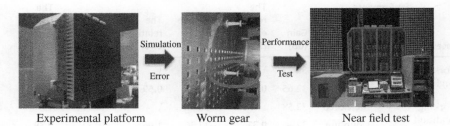

Experimental platform	Worm gear	Near field test

Figure 3.28 Physical view of the mounting and performance test environment forcing errors in the antenna.

Figure 3.29 Physical diagram of testing the electrical performance of the antenna.

by laser ranges, and the electrical properties were measured in a microwave dark-room at the State Key Laboratory of Antennas and Microwave Technology, Nanjing Institute of Electronics, China.

3.4.6.5 Calculated and Measured Results

Table 3.5 lists the calculated and measured results and comparisons for three different operating conditions. For each condition, the calculated results using the electromechanical coupling theory model are listed separately in comparison to

Table 3.5 Analysis and experimental results of an active phased array antenna with frequency of 10 GHz.

Cases	Performance	Gain (dB)	The relative error (%)	Sidelobe level (dB)	The relative error (%)	3 dB beam width(°)	The relative error (%)
Working condition No. 1	Measured	32.73	\	−13.67	\	10.14	\
	Coupling	32.65	**0.24**	−13.75	**0.59**	10.16	**0.20**
Working condition No. 2	Measured	32.58	\	−13.22	\	10.17	\
	Coupling	32.46	**0.37**	−13.36	**1.06**	10.21	**0.39**
Working condition No. 3	Measured	32.34	\	−12.51	\	10.32	\
	Coupling	32.16	**0.56**	−12.73	**1.76**	10.40	**0.77**

the measured results. The maximum relative errors of the electromechanical coupling theory model in relation to the measured results for the three operating conditions are 0.56% gain, 1.76% maximum sidelobe level, and 0.77% 3 dB beam width. The results have demonstrated the three-field coupling theoretical model of APAA.

3.5 High-density Cabinets

High-density assembly systems are widely used in telecommunication, navigation, detection, radar, radio astronomy, and so on, of which the chassis and cabinets are typical representatives. In the analysis and design of such equipment, in addition to meeting the rigidity and strength properties of the mechanical structure, it is also necessary to satisfy the requirements of electromagnetic compatibility, electromagnetic screen effect, and other electrical performance indicators. For the sake of this, it is necessary to coordinate electromagnetic, thermal, structural, and other aspects of performance, because they are often contradictory.

The following is an example of a chassis. It is supposed that there are N_e electronic devices within the chassis of a high-density assembly system, and e_i is the electric field intensity emitted by the ith device, and the magnitude of the field strength at point P at a distance d from the center of the chassis, with and without the chassis, are $\left|\sum_{i=1}^{N_e} E_i(e_i)\right|$ and $\left|\sum_{i=1}^{N_e} E_i^0(e_i)\right|$, respectively.

Thus, the effect of the chassis electromagnetic shielding is

$$SE = 20\log \frac{\left|\sum_{i=1}^{N_e} E_i^0(e_i)\right|}{\left|\sum_{i=1}^{N_e} E_i(e_i)\right|} \tag{3.54}$$

Among them, it will be affected by various factors, such as contact gap, ventilation and heat dissipation hole, structural deformation, and heat, which will be discussed separately below [23–28].

3.5.1 Effect of Contact Gaps

Gaps are often the major factors affecting the shielding efficiency, which is mostly handled in engineering by filling them with conductive materials, and the effect of gaps on electromagnetic leakage can be derived from the method of transferring impedance.

It is assumed that the thickness of the gap is W_1, and the contact area of the gap S = stitch length × stitch width. If the thickness of the filling material is W_2, and the gap transferring impedance of the gap is Z, the conductivity of the filling material will be $\sigma = \frac{\sqrt{W_1 W_2}}{SZ}$. Since the electricity of the shielded body with gap is the function of transferring impedance Z, which is the function of frequency, the denominator of formula (3.54) will become

$$\left| \sum_{i=1}^{N_e} E_i(e_i, Z(freq)) \right| \tag{3.55}$$

3.5.2 Effect of Heat Sink Holes and Structural Deformation

As shown in Figure 3.30, the chassis has electromagnetic emitting components v_i and v_j, circular and rectangular holes for ventilation and heat dissipation, and the structural parameter β of the chassis includes the wall thickness of the chassis, the size of the reinforcement, the thickness of the internal partition, the size and location of the heat dissipation holes, and the location of the internal electronic components. When the chassis is subjected to external loads P_1 and P_2, the structure of the chassis will theoretically be deformed, which will not only cause a change in the location of the internal components but also lead to a change in the distribution of the electromagnetic and temperature fields inside the chassis. As a result, the electromagnetic shielding effect of the chassis will be changed.

The reason for the above can be seen as follows. When the chassis is deformed under loads, the direction of the current vector at its induction surface will be changed, as shown in Figure 3.31, and the vector direction can be expressed in terms of the derivative of the structural deformation δ. The deformation does not change the amplitude or phase of the current. Meanwhile, the structural deformation $\delta(\beta)$ is the function of the structural parameters β.

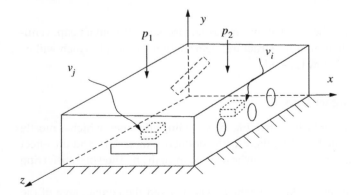

Figure 3.30 The chassis under loads.

Figure 3.31 Surface currents before and after deformation. (a) Undeformed surface current. (b) Deformed surface current.

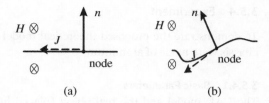

(a)   (b)

In addition, changes in the temperature field will also cause changes in the displacement and electromagnetic fields, as changes in temperature will cause changes in the electrical properties of the electromagnetic parts (devices) in addition to structural deformation. The two factors can be expressed mathematically as $\delta(\beta, T)$ and $\left| \sum_{i=1}^{N_e} E_i(e_i(T)) \right|$, respectively.

Taking the above factors into account, Eq. (3.55) can thus be further written as

$$\sum_{i=1}^{N_e} E_i(e_i(T), Z(freq), \delta(\beta, T)) \tag{3.56}$$

3.5.3 Theoretical Model of Electromagnetic–Displacement–Temperature Fields Coupling

The analysis above tells us that the high-density chassis has the mutual influence and coupling relationship among the structural displacement field, electromagnetic field, and temperature field. And the coupling relationship will affect the electromagnetic shielding effect of the chassis; therefore, the establishment of electromagnetic–displacement–temperature fields coupling model is a critical issue, for only in this way can we fundamentally reveal their mutual influence law and guide the engineering practice.

In summary, the theoretical model of the electromechanical thermal field coupling of the high-density chassis can be mathematically represented as

$$SE = 20 \log \frac{\left| \sum_{i=1}^{N_e} E_i{}^0(e_i) \right|}{\left| \sum_{i=1}^{N_e} E_i(e_i(T), Z(freq), \delta(\beta, T)) \right|} \tag{3.57}$$

Based on this coupling model of mechanical, electromagnetic, and thermal fields, a multidisciplinary coupling optimization design model can be established to guide the engineering design.

3.5.4 Experiment

To demonstrate the proposed theoretical model of the high-density chassis, two chassis were made of aluminum.

3.5.4.1 Basic Parameters

The CAD model and the real object (placed horizontally) of chassis No. 1 are shown in Figures 3.32 and 3.33, respectively, with an external dimension of $401 \times 171.5 \times 162$ mm. The chassis is used to verify the effect of structural parameters on electrical performance and has interchangeable front side panels in four different opening sizes (Figure 3.34). As the field source is horizontally polarized during the simulation, the width of the hole has little effect on the screen effect, while the length of the hole has a greater effect. For this reason, the hole width is set to be 8 mm, while the hole lengths are 10, 20, 30, and 40 mm, respectively. The remaining cover panels have no holes and are bolted together, and the bolt spacing is less than 70 mm.

The CAD model and the physical object (chassis placed horizontally) of chassis No. 2 are shown in Figures 3.35 and 3.36, respectively, with external dimensions of $320 \times 330 \times 400$ mm^3, and it is made of aluminum alloy. As shown in Figure 3.37, chassis No. 2 was used to verify the effect of contact gaps on electrical performance.

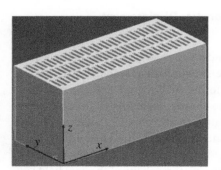

Figure 3.32 Diagram of chassis No. 1.

Figure 3.33 Physical view of chassis No. 1.

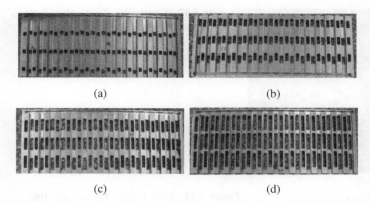

Figure 3.34 Physical view of the four opening sizes of the cover plate. (a) 10*8. (b) 20*8. (c) 30*8. (d) 40*8.

Figure 3.35 Diagram of chassis No. 2.

Figure 3.36 Physical view of chassis No. 2.

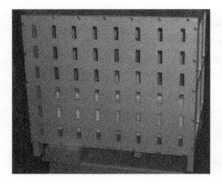

Figure 3.37 Chassis No. 2 with open front panel installed.

Figure 3.38 Model of the chassis with the conductive rubber installed.

The chassis is equipped with two front panels, one without a hole and the other with an open hole of $30 \times 10 \, \text{mm}^2$. The front panel has mounting slots for the conductive rubber, which is mounted in the positions shown in the highlighted lines in Figure 3.38. The bolts are spaced less than 70 mm apart, and all parts are connected by welding except the front panel.

3.5.4.2 Measurement and Environment

The test block diagram for this experiment is shown in Figure 3.39, and $d = 1 \, \text{m}$. The experimental site is located in a darkroom with 5 m of an EMC research laboratory (Figure 3.40).

The experimental apparatus consists of a broadband electromagnetic signal source (Figure 3.41) and a receiving antenna (Figure 3.42). The signal source is a spherical dipole antenna with a diameter of 100 mm, and its frequency range is 30–990 MHz. With a frequency interval of 30 MHz for the experiment, a total of 33 test frequencies are applied.

Electronic cabinet under test

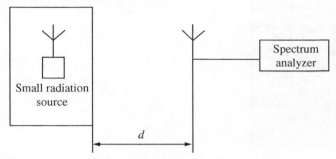

Figure 3.39 Block diagram of chassis shielding effectiveness test.

Figure 3.40 Darkroom test site with 5 m.

Figure 3.41 Wideband signal source.

Figure 3.42 Receiving antenna.

3.5.4.3 Calculated and Measured Results

Influence of Structural Parameters on Electrical Properties For chassis No. 1, the impact of structural parameters such as hole gaps on the electrical performance of the chassis, including shielding effectiveness, was verified by replacing the front covering plate with a different opening size. Step laps were designed at the contact gaps of each covering plate in chassis No. 1 to increase the gap depth, while the bolt spacing did not exceed 70 mm. Both measures reduced the degree of electromagnetic field leakage through the contact gap effectively. At the same time, as the size of the openings is much larger than the contact gaps, the main electromagnetic leakage path for this chassis is the openings in the front covering plate.

The simulated and measured results for the four openings are shown in the four diagrams in Figure 3.43, with average errors of 6.08, 5.49, 3.60, and 4.76 dB, respectively, all of which are below the 10 dB margin of error generally allowed in engineering.

Effect of Contact Gaps (Conductive Rubber) on Electrical Properties The equivalent model of the contact gap based on transferring impedance in the triple-field coupling model was verified with chassis No. 2, and conductive rubber studied. The conductive rubber was compressed to a height of 2.5 mm. Figure 3.44 provides a comparison of the simulated and measured results of the electromagnetic shielding effect of chassis No. 2 when the conductive rubber is installed. It can be seen that the trend between the test and simulation is very close, with an average error of 3.48 dB.

Consideration About the Impact of Four Factors on Electrical Performance The experiments above just consider the influence of a single major factor, whereas in actual engineering there are four factors at the same time: structural parameters, contact

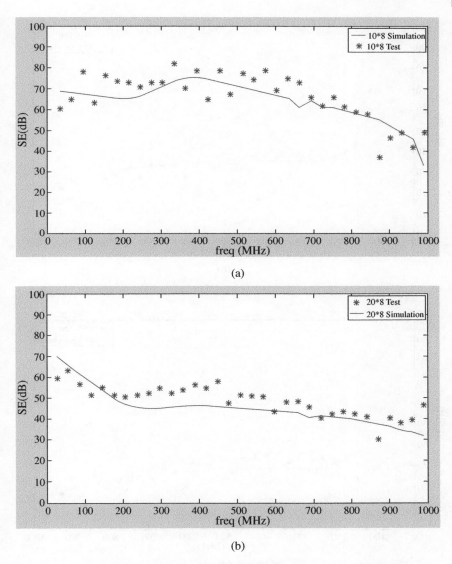

Figure 3.43 Test and simulation results of the shielding effectiveness of chassis No. 1 with four openings (the discrete points on the graph are the testing results). (a) Opening 10*8. (b) Opening 20*8. (c) Opening 30*8. (d) Opening 40*8.

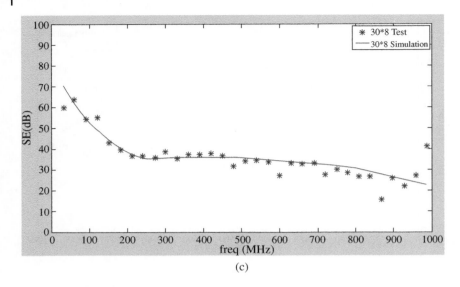

(c)

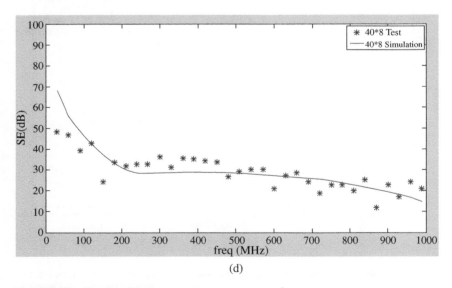

(d)

Figure 3.43 (*Continued*)

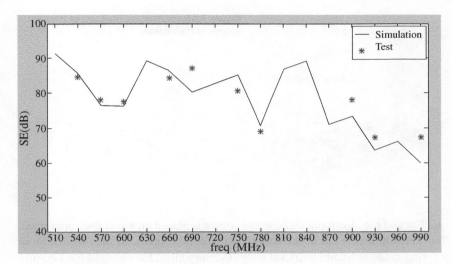

Figure 3.44 Conductive rubber shielding effectiveness test and simulation results for chassis No. 2.

gaps, structural deformation, and temperature. For this reason, an experimental demonstration including these four factors was also carried out.

The influence of structural parameters can be achieved by installing openings in the front panel of chassis No. 2. The influence of contact gaps can be achieved by the installation of conductive rubber in chassis No. 2. The effect of structural deformation and temperature can be achieved by means of a specially designed bracket structure which is placed inside the chassis shown in Figure 3.45. The stand (in order not to affect the electromagnetic field distribution inside, the bracket structure is made of wood) consists of an upper plate, a floor, spacers, and springs. The floor is connected to the bottom of the chassis so that heat can be transferred up

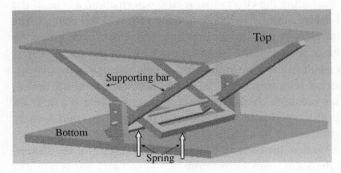

Figure 3.45 Schematic diagram of the internal bracket structure of the chassis.

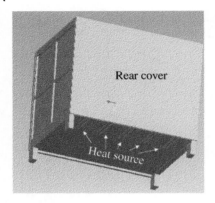

Figure 3.46 Schematic diagram of the heat source of the chassis.

quickly, and the electromagnetic signal source is placed on top of the upper plate. Two small holes are cut in the bottom plate of the support for the installation of a spring made of titanium–nickel alloy (memory alloy, the changed temperature of the phase is 70 °C). At room temperature, the spring contract and the bracket spacers are in direct contact with the bottom plate. When the temperature inside the chassis rises above 70 °C, the spring begins to elongate, lifting the spreader bar up and causing the signal source placed on the upper plate of the bracket to drop 2 cm to reflect the idea that the temperature's change can cause a change in the position of the internal signal source.

For this purpose, the heat source shown in Figure 3.46 is installed on the base of the chassis and is made of 10 soldering iron cores with the power 60 W. At the same time, temperature sensors are placed in several typical locations inside the chassis in order to measure the internal temperature distribution in time. Note that when heating underneath, it will cause a change in the length of the spring placed in the bottom bracket of the heated chassis, which in turn causes a change in the position (downward shift) of the signal source placed in the upper plate of this bracket, demonstrating the effect of structural deformation, as one of the results of structural deformation is a change in the position of the components and devices inside the chassis. In addition, the change in temperature inside the chassis will also cause a change in the electrical properties of the signal source. Both will ultimately result in a change in the leakage field strength of the chassis.

Regarding the effect of temperature on the electrical properties of the device, Figure 3.47 depicts the radiation characteristics of the signal source as a function of temperature (7 °C [the experiment was conducted in Nanjing in mid-January 2011] room temperature and 45 °C high temperature).

Figures 3.48 and 3.49 show the measured and simulated results for the leakage field strength of the chassis at room temperature of 7 °C and high temperature of 45 °C, with an average error of 3.79 and 3.44 dB, respectively, both below 10 dB

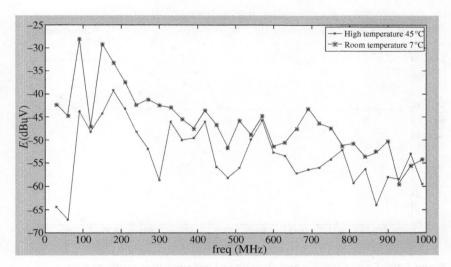

Figure 3.47 Variation of the radiation characteristics of the signal source with temperature.

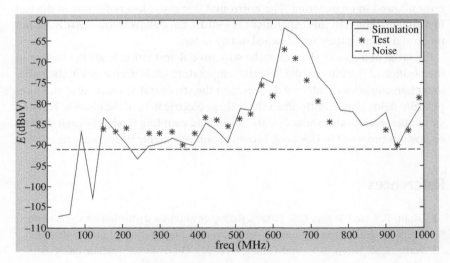

Figure 3.48 Test and simulation results of chassis leakage field strength at room temperature.

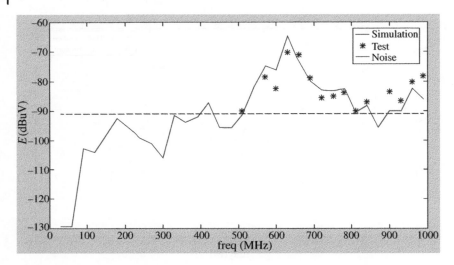

Figure 3.49 Chassis leakage field strength test and simulation results at high temperature.

error allowed in engineering. The horizontal line is the low noise line of the test instrument, which in this experiment is −90 dB, i.e. signals below −90 dB are not measured because they are drowned out by noise.

Experimental results show that the structural deformation leads to changes in the position of the internal devices, the temperature leads to changes in the radiation characteristics of internal devices, and the structural displacement and temperature fields significantly affect the leakage electric field of the chassis. Test and simulation results also show that the three-field coupling model for high-density enclosures proposed in this book is correct and acceptable in engineering.

References

1 Duan, B.Y. and Wang, C.S. (2009). Reflector antenna distortion analysis using MEFCM. *IEEE Transactions on Antennas and Propagation* 57 (10): 3409–3413.

2 Wang, C.S., Duan, B.Y., and Qiu, Y.Y. (2007). On distorted surface analysis and multidisciplinary structural optimization of large reflector antennas. *International Journal of Structural and Multidisciplinary Optimization.* 33 (6): 519–528.

3 Wang C S, Duan B Y, Qiu Y Y, et al. (2008). On coupled structural-electromagnetic optimization and analysis of large reflector antennas. *The 8th International Conference on Frontiers of Design and Manufacturing (ICFDM2008)*, September 23–26, 2008, Tianjin, China.

4 Wang, C.S. and Bao, H. (2008). Analysis and coupling optimization design of intelligent antenna structural systems in satellite. *The 26th AIAA International Communications Satellite Systems Conference (ICSSC 2008)*, June 10–12, 2008, San Diego, CA, AIAA-2008-5485.

5 Wang, C.S., Bao, H., and Wang, W. (2008). Coupled structural-electromagnetic optimization and analysis of space intelligent antenna structural systems. *The 9th Biennial ASME Conference on Engineering Systems Design and Analysis (ESDA08)*, July 7–9, 2008, Haifa, Israel.

6 Cong-Si, W.A.N.G., Bao-Yan, D.U.A.N., Fei, Z.H.E.N.G., and Yuan-Ying, Q.I.U. (2008). Mechatronic optimization design and analysis of large space parabolic antennas with active truss support structures. *Acta Electronica Sinica* 36 (9): 1776–1781.(in Chinese).

7 Cong-Si, W.A.N.G., Bao-Yan, D.U.A.N., and Yuan-Ying, Q.I.U. (2008). On new fitting method of large distorted antenna reflectors based on coons surface and B-spline. *Journal of Electronics & Information Technology* 30 (1): 233–237. (in Chinese).

8 Li-Wei, S., Bao-Yan, D., Fei, Z., and Hong-Bo, M.A. (2009). The effects of surface error on reflector antenna performance. *Acta Electronica Sinica* 37 (3): 552–556. (in Chinese).

9 Li-Wei, S. and Fei, Z. (2009). Analysis of the coupled problems between structure and electromagnetism based on discrete meshes. *Journal of Xidian University* 36 (2): 347–352. (in Chinese).

10 Li-Wei, S.O.N.G., Bao-Yan, D.U.A.N., and Fei, Z.H.E.N.G. (2009). Effects of reflector error and phase center errors of feed on the far field pattern of reflector antennas. *Systems Engineering and Electronics* 31 (6): 1269–1274. (in Chinese).

11 Li-Wei, S.O.N.G., Bao-Yan, D.U.A.N., and Fei, Z.H.E.N.G. (2009). A study on the optimal FEEDS phase Center for distorted reflector antennas. *Transactions of Beijing Institute of Technology* 29 (10): 894–897. (in Chinese).

12 Song, L.W. (2010). Performance of planar slotted waveguide arrays with surface distortion. *Progress In Electromagnetic Research Symposium 2010*, March 22–26, Xi'an, China.

13 Song, L.W. (2010). Analysis of integrated structure-electromagnetic wave basing on the same discrete meshes. *Progress In Electromagnetic Research Symposium 2010*, March 22–26, Xi'an, China.

14 Guo-Qiang, C.H.E.N. and Min-Bo, Z.H.U. (2008). Thermal test numerical simulation on heat transfer characteristics of forced air cooling for electronic equipment. *Computer Aided Engineering* 17 (2): 24–26. (in Chinese).

15 Xin-an, L.I., Hua, J.I.A.N.G., Fu-shun, Z.H.A.N.G., and etc. (2007). Summary of electromechanical thermal field coupling research. Journal of. *Radio Science* 22 (S): 31–34. (in Chinese).

16 Wang, C.S. and Duan, B.Y. (2010). Coupled structural-electromagnetic-thermal modeling and analysis of active phased array antennas. *IET Microwaves, Antennas & Propagation* 4 (2): 247–257.

17 Wang, C.S., Duan, B.Y., Zhang, F.S. et al. (2009). Analysis of performance of active phased array antennas with distorted plane error. *International Journal of Electronics* 96 (5): 549–559.

18 Cong-Si, W., Li-Hao, P., Meng, W., and Hui-Juan, X. On analysis of distorted phased Array antennas based element coupling. *10th National Radar Annual Conference*, 2009.10.14–16, Sichuan, Chengdu. (in Chinese).

19 Cong-Si, W., Xin-An, L., Fu-Shun, Z., et al. Modeling and analysis of structure and electromagnetic coupling of rectangular active phased Array antenna. *10th National Radar Annual Conference*, 2008.10.31–11.2, Beijing. (in Chinese).

20 Ge, C., Duan, B., Wang, W. et al. (2019). An equivalent circuit model of the rectangular waveguide performance analysis considering rough flanges' contact. *IEEE Transactions on Microwave Theory and Techniques* 67 (4): 1336–1345.

21 Lou, S., Duan, B., Wang, W. et al. (2019). Analysis of finite antenna arrays using the characteristic modes of isolated radiating elements. *IEEE Transactions on Antennas and Propagation* 67 (3): 1582–1589.

22 Duan, B. (2020). The scientific basis of functional surface precise design and performance guarantee[R], Summary Report on the major projects of the National Natural Science Foundation of China.

23 Duan, B.Y. and Wang, C.S. Analysis and optimization design of multi-field coupling problem in electronic equipments. In: *International Workshop 2007: Advancements in Design Optimization of Materials, Structures and Mechanical Systems*, 17–20 December 2007, Xi'an, China, 252–261.

24 Duan, B.Y. (2010). The multi-field-coupled model and optimization of absorbing material's position and size of electronic equipments. *Journal of Mechatronics and Applications* 1–6.

25 Guo-Qiang, S. (2009). Design of stiffener in the electronic cabinet based on the three-field coupling analysis [Dissertation]. Xidian University. (in Chinese).

26 Yu, H. (2009). Thermal design of components placement based on three-field-coupling in electronic equipment [Dissertation]. Xidian University. (in Chinese).

27 Hui, Q. (2009). Optimization of absorbing material attaching based on three-field-coupling in electronic equipment [Dissertation]. Xidian University. (in Chinese).

28 Shi-Bo, J. (2009). Research on three-field-coupling analysis of electronic equipments [Dissertation]. Xidian University. (in Chinese).

4

Solving Strategy and Method of the Multifield Coupling Problem of Electronic Equipment

4.1 Introduction

Aiming at solving the multifield coupling model of the electromagnetic, structural displacement, and thermal fields established in Chapters 2 and 3, this chapter considers the solution problem. To do this, the characteristics of each physical field are analyzed first, and the mesh of each physical field used in numerical calculation and information transformation among different meshes are studied. On the basis of the research on the solving method of field coupling problem, the field coupling model calculation program is developed by comprehensively utilizing the existing software tools of calculating structural displacement and electromagnetic and temperature behaviors, which lays a foundation for the development of field coupling comprehensive design software platform in the following chapters of this book.

4.2 Solving Strategy of the Multifield Coupling Problem

There are two kinds of solving strategies for CMFP, that is analytical solving strategy (ASS) and comprehensive solving strategy (CSS).

The ASS is that each physical field establishes its mathematical model separately, and the analysis of each field model is connected through boundary conditions or other coupling relations, and then the solution strategy is carried out. In this book the ASS is employed. There are four main kinds of ASS, including direct coupling, sequential coupling, mathematical decoupling, and comprehensive optimization analysis. Direct coupling analysis is to analyze the mathematical model of CMFP as a whole, so that the analysis results of each physical field can be obtained simultaneously. Sequential coupling analysis is an analysis in which two or more physical fields are arranged in a certain order for solution, and the

Electromechanical Coupling Theory, Methodology and Applications for High-Performance Microwave Equipment, First Edition. Baoyan Duan and Shuxin Zhang.

coupling is performed by applying the analysis results of the previous physical field as a load to the analysis of the next physical field. Mathematical decoupling analysis is to decouple the coupling equations by mathematical methods and obtain equations with independent variables. Since the fields are interrelated, even though the analysis can be performed independently, the final result must be obtained by iterations among them until the convergence, where the final result of the individual physical quantities can be obtained. Comprehensive optimization analysis is an analysis method to solve the mathematical model of CMFP by adopting multidisciplinary optimization strategies and methods, which can solve the difficulties of CMFP complex calculation. This strategy analyzes the influence behavior of each physical field under the coupling action and reflects the local characteristics of the influence relationship among different physical fields.

The comprehensive solution strategy is a strategy that builds a macroscopic model of the CMFP system based on the analysis of each physical field and the coupling relationship with appropriate simplification of the coupling fields and then solves it. This strategy is usually used for system-level analysis studies because system-level analysis does not deliberately pursue local characteristics but places great emphasis on the overall system performance instead, i.e. the input–output characteristics. It is suitable for the analysis of MEMS (microelectromechanical system). Establishing a macroscopic model of a MEMS system is the key to system-level analysis, and there are two main approaches [1], the nodal analysis method and the black box method. The former one considers the system as composed of several basic elements, each element is a node, and uses the analog hardware description language to connect the above nodes with the real circuit to form a network and establishes the differential equations of the system. The latter selects a few parameters to describe each physical field, turns the coupling field into a black box described by a few parameters, and then inserts the black box into the MEMS circuit and applies circuit analysis software to it. This solution strategy lies in the analysis of the overall characteristics of the coupling field, which reflects the overall system performance of the coupled field.

4.3 Solving Method of the Multifield Coupling Problem

The solving method of CMFP mathematical model given in this book adopts an analytical solution strategy, and the corresponding analytical and solution methods are shown in Table 4.1.

4.3.1 Solution Method of Direct Coupling Analysis

The solution method is an integrated solution method [2, 3] (direct method, full coupling method), i.e. the equations of all physical fields are solved simultaneously

Table 4.1 Solving strategies and methods for CMFP models.

	Solution strategy	Analysis method	Solving method
CMFP models	Analysis solving strategies (ASS)	Direct coupling	Integration
			Direct solution
			Fully coupled solution
		Sequential coupling	Iteration
			Partitioned solution
			Domain decomposition
		Mathematical decoupling	Mathematical transformation
		Comprehensive optimization	Optimized solution
	Comprehensive solving strategy (CSS)	Nodal analysis	Differential equation
		Black box method	Circuit analysis

in the same step. The coupling relations can be solved with each physical field either synchronously or asynchronously. Theoretically, the integrated solution method is applicable to highly nonlinear coupling problems. It needs to give its solution algorithm for specific coupling problems, so it lacks generality.

4.3.2 Solution Method of Sequential Coupling Analysis

The solution method of sequential coupling analysis is a partitioning method [4, 5] (indirect method, iterative method) in which each field is solved sequentially in one step, and the coupling information is passed between fields. The partitioning method is suitable for coupling problems without a high degree of nonlinearity. Since the partition method has the advantages of separate modeling, using respective simulation tools, ensuring software reuse, and performing modularization, it is widely used in the solution of coupled problems in the field. The shortcomings of the partitioning method are slower convergence, less stability, and lower accuracy than the integrated solution method, which is especially clear for the time-varying CMFP.

4.3.3 Solution Method for Mathematical Decoupling Analysis

The solution method of mathematical decoupling analysis is a method based on the basic assumption that the system of coupled equations can be decoupled, which is similar to the solution method of solving single field and will not be discussed further. Note that the difficulty of this approach lies in the mathematical decoupling. However, not all systems of coupled equations can be decoupled in practical problems.

4.3.4 Solution Method of Integrated Optimization Analysis

Integrated optimization is also one of the effective methods to solve CMFP. In particular, the multidisciplinary optimization method is applied to the conceptual design of aircraft in the field of aerospace. The multidisciplinary optimization approach is one of the effective methods to solve the optimal design of large complex engineering systems of aircraft type. It should be noted that in applying the multidisciplinary optimization [6] method to solve CMFP problems, the fast reanalysis of each nonlinear problem is fundamental, and one of the approaches to achieve fast reanalysis is the surrogate model technique, i.e. to speed up the iteration convergence while ensuring the accuracy requirements. The commonly used surrogate models are the polynomial response surface model, the radial basis function model, and the Kriging model.

The flow of fluid–solid coupling analysis procedure [7] is given in Figure 4.1, which is a time-varying partitioning method analysis procedure. The fluid analysis and structural analysis are carried out by choosing the numerical solution methods suitable for their own fields, and this part of the work can be carried out by commercial software. The coupling information is transferred through the interface equilibrium condition. Within each step, the pressure information obtained from the fluid analysis is matched to the structural mesh nodes to form the loads required for the structural analysis after which the displacement information of the structure is obtained by the structural analysis. As the next step of the iteration, the displacement information is matched to the fluid mesh nodes to form the adjusted fluid computational mesh with boundary information. If both fluid and structure are analyzed simultaneously, and the obtained results with coupled information are brought to the next step for solution, this is the integrated solution method.

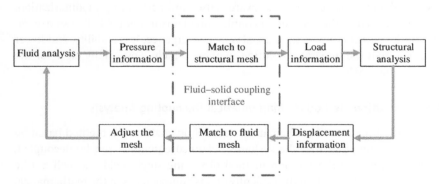

Figure 4.1 Analysis flow of fluid–solid coupling problem.

4.4 General Approach Method of the Multifield Coupling Problem

From the above analysis, it is clear that the coupling information transformation between the physical field meshes is the key to perform the coupling analysis. First, it should be done to understand the relationship between the different physical field meshes. For different forms and accuracy, the different physical field meshing will not only introduce the mismatch between the meshes of different physical field common interfaces but also bring mesh gaps and mesh coverage [8]. Figure 4.2 gives a two-dimensional case of mesh mismatch for discrete interfaces of different physical fields. In Figure 4.2, Γ is the continuous common boundary between different physical fields, and Γ_A, Γ_B are the discrete boundaries of different physical fields. The information on the discrete boundaries Γ_A, Γ_B of different physical fields is related by mathematical physics to establish the information transformation.

For coupling information transformation in field coupling problems, there are several criteria that such a data exchange or coupling method ideally should be satisfied. The most important aspects are [8]: (i) global conservation of energy over the interface, (ii) global conservation of loads over the interface, (iii) accuracy, (iv) the conservation of the order of the coupled solvers, and (v) efficiency, which is defined as the ratio between accuracy and computational costs.

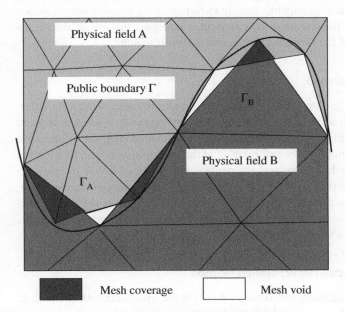

Figure 4.2 Mismatching mesh between physical fields.

Assuming that the coupling information at the discrete interface of the abovementioned two-dimensional physical fields is the same as that described in the mathematical model of the field coupling problem in the general sense, the coupling information needs to be transformed to satisfy the requirements of each physical field due to $\Gamma_A \neq \Gamma_B \neq \Gamma$, such that Eqs. (4.1) and (4.2) can be transformed into

$$Y_{A \to B} = H_{A \to B} X_{A \to B} \tag{4.1}$$

$$X_{B \to A} = H_{B \to A} Y_{B \to A} \tag{4.2}$$

where $H_{A \to B}$ and $H_{B \to A}$ are transformation matrices of the coupling information transformation, and $X_{A \to B}$, $Y_{A \to B}$, $X_{B \to A}$, and $Y_{B \to A}$ are the information vectors of the interaction between the physical fields, respectively. As for how to determine the conversion matrix, it can be derived from the transformation relation given above. Some common methods of determining the conversion matrix are as follows.

4.4.1 Neighborhood Interpolation Method

The neighborhood interpolation method [9] is a convenient interpolation method. The basic idea is to determine a node x_B in mesh B that is closest to a node in mesh A by a search algorithm and then pass the information at x_A directly to x_B. In this way, only one item in each row of the transformation matrix H is equal to 1, and the rest are 0. This method can only obtain satisfactory results when the meshes of the physical fields A and B almost match.

4.4.2 Mapping Method

In order to be able to obtain the information of nodes in mesh A from the information of mesh B, it is necessary to map node x_A orthogonally to a mapping point of mesh B. The coupling information of node x_A is obtained from the information of the mapping point [10]. The information of the mapping point can be obtained by interpolation method. Similarly, elements can be mapped by the intersection method [11]. Due to the use of orthogonal mapping, some nodes or elements cannot be mapped to the corresponding mesh, which can easily cause the corresponding coupling transformation imbalance. Therefore, a compensation method needs to be considered to solve this problem.

4.4.3 Spline Function Interpolation Method

The use of the spline basis functions to describe the displacements of the fluid and structural meshes is to establish the transformation of information between the two meshes by obtaining the transformation matrix H through the interface equilibrium condition. In the literature [12], C^2 radial basis functions are applied to establish the transformation matrix of coupling information in fluid–solid coupling problems. In the literature [13], the quadratic surface bi-tuning spline function and the thin plate spline function are applied in realizing the establishment.

4.4.4 Continuation Method

This is a method to transfer information by extending the element in the usual sense without increasing the number of nodes [14]. It first defines the nested element domain and uses the nodal information outside the element to force the interpolation function inside the element to be of higher order, thus constructing a generalized shape function and finally obtaining a high-precision method for constructing a mesh shift using the information from the outer points.

The information transformation methods mentioned above obey the law of energy conservation at the interface and the condition of interfacial force equilibrium. The analytical errors come from interpolation and mapping. In terms of accuracy and computational efficiency, the spline function method is more superior. The high-precision continuation method will also be developed with the improvement in computational accuracy.

It can be seen that the key to the accuracy of the CMFP model solution lies in the accuracy of coupling information transformation. With the development of electronic equipment of high-frequency band, high gain, fast response, high pointing accuracy, high density, and miniaturization, the field coupling theory model among electromagnetic field, displacement field, and temperature field will become more complex, and the accuracy and completeness of the information transferred among the fields will be more and more demanded, so the research of an efficient and high-accuracy solution method is highly urgent.

4.5 The Mesh Matching Among Different Fields

The meshes of structural finite element model and the EM field calculation and the conversion and transformation of information include the following two ways [15–18]: one is the EM mesh generated directly in the already established structural finite element mesh, and the other is generated through the mesh-mapping approach.

4.5.1 Generated Directly in the Structural Finite Element Mesh

As shown in Figure 4.3, taking the midpoints of the three sides and connecting them to create four triangles within triangle ABC, and obeying this order,

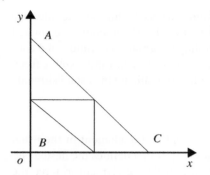

Figure 4.3 EM mesh generation directly in the structural triangular element.

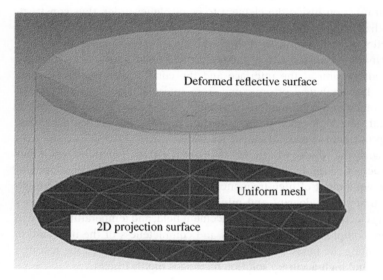

Figure 4.4 Mesh mapping between reflective surface structure and EM.

generating 16, 64, and 256 triangles until the principle that the lengths of the sides of the triangles do not exceed the 1/5–1/8 of working wavelength.

4.5.2 Mesh Mapping from Structure to EM

Taking the large reflector antenna as an example, the inconsistency of the mesh formed by structural analysis and electromagnetic analysis and the large scale of electromagnetic model make the coupled structural–electromagnetic analysis very difficult. For the sake of this, the function mapping idea is applied to the mesh information conversion of the structural analysis model and the electromagnetic analysis model (Figure 4.4), which is implemented in the following basic steps:

(1) The reflector antenna is divided into finite element mesh for structural analysis, and the deformed reflector surface 3D structural mesh is obtained.

(2) On the two-dimensional projection plane of the original reflective surface, the mesh is divided according to the principle of working wavelength 1/5–1/8 to form a relatively uniform planar triangular mesh.

(3) A uniform planar triangular mesh is mapped onto the deformed reflective surface, and a 3D electromagnetic mesh with sampled deformation information is obtained by interpolation. The mesh is relatively homogeneous and satisfies the electromagnetic calculation requirements, thus enabling structural–electromagnetic analysis.

4.6 Mesh Transformation and Information Transfer

High-density chassis or active phased array antennas involve the calculation of electromagnetic fields, displacement fields, and temperature fields, which require the transformation of coupling information among three physical fields. The most common procedure is to transfer the structural deformation information to the temperature analysis module and the electromagnetic analysis module, respectively. Although the physical equations in the three disciplines are different, the numerical analysis methods are all performed on the discretized meshes and importantly allow easy transformation of the coupling information. In fact, structural analysis mostly uses finite element method; electromagnetic analysis employs finite element method, time domain finite difference method, and moment of method; and temperature field analysis mostly uses finite volume method. The three physical fields have not only different analysis methods but also different mesh forms. The structure has various element types and different mesh forms, such as the widely used triangular, quadrilateral, tetrahedral, hexahedral, and other elements. The finite element method and the method of moments in electromagnetic analysis are generally performed by tetrahedral and triangular elements, while the time domain finite difference method uses hexahedral and quadrilateral meshes. Hexahedral elements are commonly used in temperature analysis. Obviously, it is very difficult to unify the mesh of the three physical fields in a broad sense.

The transfer of deformation information through a set of discrete meshes is still possible if the specific algorithm for each physical field is determined. The Pro/E software is used in parametric modeling, ANSYS and Workbench are used in structural analysis, FEKO is used in electromagnetic analysis, and ICEPAK software is used in temperature field analysis. Application of these commercial software for solid modeling and multiphysics field numerical analysis is the basis, but the key to multifield coupling analysis and design is the need for the secondary development, namely, the establishment of electromechanical–thermal coupling mathematical model, the development of the corresponding professional software, and the formation of the electromechanical–thermal coupling design platform for electronic equipment, which is a precisely important and difficult work.

4.6.1 Transmission of Deformation Information

When performing structural finite element analysis, the chassis structure often uses plate, shell, and body elements. The plate and shell elements are mainly triangular and quadrilateral plate and shell elements, while the body elements are mainly tetrahedral elements and hexahedral elements. ICEPAK software for temperature field calculation is able to import models that are assembled with planar triangular elements. In this case, the geometry formed by the triangular elements is treated as a solid, on which the respective mesh is divided, and the analysis is completed. Therefore, by extracting the nodal deformation coordinates and element information on the outer surface of the structural mesh, and reorganizing the elements to form triangular slices, and then writing them into a mesh file type that can be recognized by FEKO software and ICEPAK software, respectively, the structural deformation information can be transferred to the electromagnetic analysis and temperature analysis to realize multifield coupling analysis. The basic idea of this process is to approximate the deformed surface with small triangles in the plane, thus enabling the transformation of deformation information [19, 20].

The overall flow of information transformation from structural displacement field to temperature field and electromagnetic field is shown in Figure 4.5. First, the parametric CAD solid model is imported into the structural analysis software for structural analysis, and then the deformed thermal analysis mesh and electromagnetic analysis mesh are extracted for thermal and electromagnetic analysis.

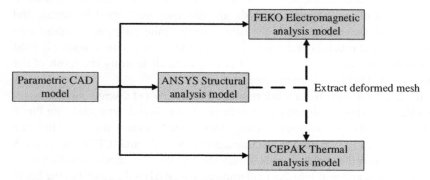

Figure 4.5 General flow of information transformation of structural displacement field.

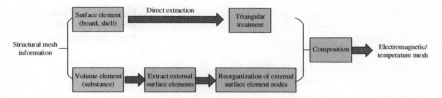

Figure 4.6 General flow of information transformation of structural displacement field.

4.6.2 Extraction of Deformed Meshes

The basic flow of deformation mesh extraction is shown in Figure 4.6, i.e. the output contains the coordinates, displacements, and element information of all nodes after the structural analysis is completed. They are handled according to the different types of elements as follows: for two-dimensional surface elements (plate and shell elements), determining whether they are triangular elements; if so, the element information is extracted directly, and if not, the triangulation process is performed and then extracted. For 3D solid elements, it is necessary to materialize into a plane first, extract the outer surface elements, delete the internal nodes and elements, and then apply the processing of 2D surface elements to proceed. Finally, the processed elements are reassembled, numbered, and written in a file format that FEKO and ICEPAK can recognize them.

The surface mesh extraction and triangulation process of the plate and shell is shown in Figure 4.7. Firstly, whether or not the plate and shell elements exist is judged, and if they do not exist, the surface mesh need not to be extracted. If it does, the plate and shell surfaces are selected to be extracted. Provided that the number of plate and shell surfaces is N_b, the surface mesh extraction and triangulation for each plate and shell surface is performed, and finally, the processed nodal element information is outputted as required by the EM analysis or thermal analysis software format.

The solid outer surface mesh extraction process is shown in Figure 4.8. First, whether the entity exists or not is determined; if it does not exist, the surface mesh is not extracted. If it exists, the extracted entities are selected, and the number of entities is set to N_t. Then, the surface nodal extraction and triangulation element reorganization process is performed for each entity. If the entity is divided into tetrahedral mesh, the surface mesh is triangular, and the surface mesh can be extracted directly. However, when the entity is divided into hexahedral mesh, the surface mesh should be triangulated and reorganized after extracting the surface mesh.

Among them, the procedure of selecting the outer surface of the solid element is also troublesome. The selection process is shown in Figure 4.9. First, according to the solid element information of each body element, its surface information

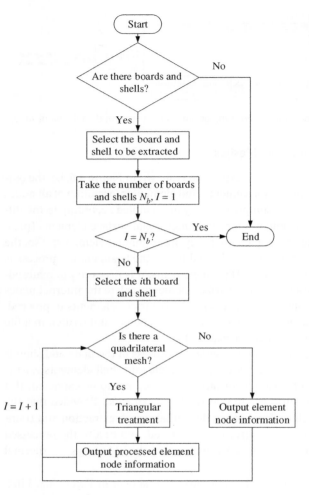

Figure 4.7 Plate and shell element surface mesh transfer flow.

is taken, such as a tetrahedral element has four faces, and a hexahedral element has six faces. It needs to be seen whether each face exists only on the same solid element; if yes, it indicates that the face is the outer surface of the element, and the face element will be saved. Otherwise, it indicates that the face is the inner surface of the element and is directly deleted, and then the next solid element is judged. After processing the complete body element in this way, the outer surface element information of a solid structure can be obtained.

Among them, the triangulation process is meant for the face element (plate and shell element). If the face element is a triangular face element, it is extracted directly. If there is a quadrilateral face element on the face, a quadrilateral should

Figure 4.8 Solid element mesh transfer flow.

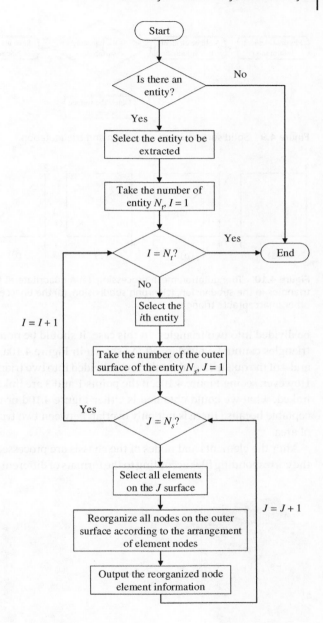

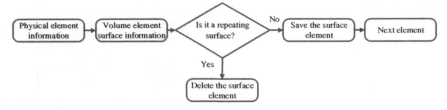

Figure 4.9 Solid surface mesh extraction and triangulation.

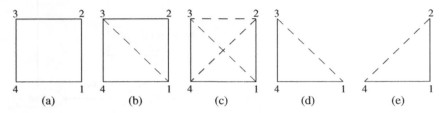

Figure 4.10 Triangulation mesh processing. (a) A quadrilateral face element, (b) two triangles in the subdivision, (c) wrong subdivision, (d) the unacceptable triangle, (e) the other unacceptable triangle.

be divided into two triangles. In this case, it should be noted that the divided two triangles cannot be overlapped. As shown in Figure 4.10a, the four nodes 1, 2, 3, and 4 of the quadrilateral mesh can be divided into two triangles as in Figure 4.10b. However, seeing Figure 4.10c, if the points 1 and 3 are linked or nodes 2 and 4 are linked, what we could obtained is either Figure 4.10d or e. It is obviously unacceptable because there is not only overlap between two triangles but also the loss of area.

After the elements and nodes of the chassis are processed, they are written into the corresponding files according to the formats of different software and imported

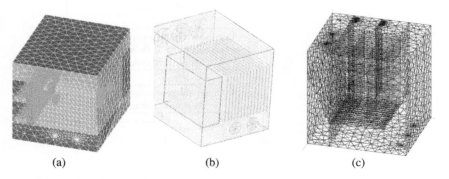

Figure 4.11 Chassis mesh conversion: (a) structural model, (b) thermal model, and (c) electromagnetic model.

into the analysis software, which completes the transformation of the mesh of the deformed structure. Figure 4.11 shows the numerical calculation mesh of the structural, electromagnetic, and thermal fields of a high-density electronic equipment chassis.

References

1 Sun, D. (2002). System-level simulation of MEMS coupled field analysis domain. *China Mechanical Engineering* 13 (9): 765–768. (in Chinese).

2 Matthias, H. (2004). An efficient solver for the fully coupled solution of large-displacement fluid–structure interaction problems. *Computer Methods in Applied Mechanics and Engineering* 193: 1–23.

3 Walhorn, E., Kolke, A., Hubner, B. et al. (2005). Fluid-structure coupling within a monolithic model involving free surface flows. *Computers and Structures* 83: 2100–2111.

4 Felippa, C. and Park, K. (2004). Synthesis tools for structural dynamics and partitioned analysis of coupled systems. *Proceedings of NATO-ARW Workshop on Multi-physics and Multiscale Computer Models in Non-linear Analysis and Optimal Design of Engineering Structures Under Extreme Conditions*, Bled, Slovenia, 6.

5 Park, Y.H. and Park, K. (2004). Anchor loss evaluation of MEMS resonators – I: Energy loss mechanism through substrate wave propagation. *Journal of Micro-electromechanical Systems* 13 (2): 238–247.

6 Jaroslaw, S. (1997). Multidisciplinary aerospace design optimization survey of recent developments. *Structural Optimization* 14 (1): 1–23.

7 Guruswamy, G.P. (2002). A review of numerical fluids/structures interface methods for computations using high-fidelity equations. *Computer Structure* 80: 31–41.

8 Boer, A.D., Van, Z.A.H., and Bijl, H. (2007). Review of coupling methods for non-matching meshes. *Computer Methods in Applied Mechanics and Engineering* 196: 1515–1525.

9 Thevenza, P., Blu, T., and Unser, M. (2000). Interpolation revisited. *IEEE Transactions on Medical Imaging* 19 (7): 739–758.

10 Cebral, J.R. and Löhner, R. (1997). Conservative load projection and tracking for fluid–structure problems. *AIAA Journal* 35 (4): 687–692.

11 Fraunhofer Institute for Algorithms and Scientific Computing SCAI (2003). MpCCI. Mesh-based parallel code coupling interface – Specification of MpCCI Version 2.0.

12 Beckert, A. and Wendland, H. (2001). Multivariate interpolation for fluid–structure interaction problems using radial basis functions. *Aerospace Science and Technology* 5 (2): 125–134.

13 Smith, M.J., Hodges, D.H., and Cesnik, C.E. (2000). Evaluation of computational algorithms suitable for fluid–structure interactions. *Journal of Aircraft* 37 (2): 282–294.

14 Jin, H., Cao, D.X., and Zhang, J. (2004). Existence and uniqueness of 2π-periodic solution about duffing equation and its numerical method. *Journal of China University of Mining & Technology* 14 (1): 104–106.

15 Rathod, H.T., Nagaraja, K.V., Venkatesudu, B. et al. (2004). Guass legendre quadrature over a triangle. *Journal of Indian Institute of Science* 84: 183–188.

16 Lague, G. and Baldur, R. (1977). Extended numerical integration method for triangular surfaces. *International Journal for Numerical Methods in Engineering* 11: 388–392.

17 Hammer, P.C., Marlowe, O.J., and Stroud, A.H. (1956). Numerical integration over simplexes and cones. *Mathematical Tables and Other Aids to Computation* 10: 130–136.

18 Chandrupatla, T. and Belegundu, A.D. (2006). *Introduction to Finite Elements in Engineering*. Beijing: Tsinghua University Press.

19 Yuqiu, L., Juxuan, L., Zhifei, L., and Song, C. (1997). Area-coordinate theory for quadrilateral elements. *Engineering Mechanics* 14 (3): 1–11. (in Chinese).

20 Zhifei, L., Juxuan, L., Song, C., and Yuqiu, L. (1997). Differential and integral formulas for area coordinates in quadrilateral element. *Engineering Mechanics* 14 (3): 12–20. (in Chinese).

5

Influence Mechanism (IM) of Nonlinear Factors of Antenna-Servo-Feeder Systems on Performance

5.1 Introduction

There are two types of electromechanical coupling problems in electronic equipment. One is the coupling problem in the form of field, and the other is the influence mechanism of mechanical structural factors on electrical performance. In the previous chapter, the modeling, analysis, and solution of the field coupling problem have been described. This chapter focuses on the analysis and discovery of the influence mechanism.

In order to deeply analyze the *influence mechanism of structure factors on performance* (named **ISFP** afterward in this chapter), two ways can be used (Figure 5.1). One is deduction, that is based on the existing data, through theoretical analysis, to find suitable design formulas, charts, and norms that can be used to guide practical engineering. Second, inductive method, that is by the successful experience and failure lessons summed up in the massive data, the use of data mining technology to find out the useful empirical formulas, charts, and design specifications. This chapter focuses on the application of the abovementioned two methods to the analysis model of the influence of structural factors on the electrical performance in the reflector antenna, the planar slotted array antenna, the resonant cavity filter, the radar antenna-servo system, and the active phased array antenna, so as to obtain the qualitative or quantitative relationship between the structural factors and the electrical performance, explore the corresponding influencing mechanism, and form the design and manufacturing specification and comprehensive design platform with engineering guiding significance.

5.2 Data Mining of ISFP

In order to effectively discover the influence mechanism of mechanical and structural factors (including structural parameters and manufacturing accuracy) on the

Electromechanical Coupling Theory, Methodology and Applications for High-Performance Microwave Equipment, First Edition. Baoyan Duan and Shuxin Zhang.

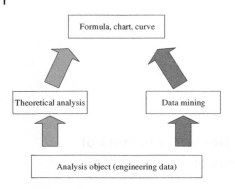

Figure 5.1 Two ways to discover the influence mechanism.

electrical performance of electronic equipment, that is to obtain implicit laws from existing engineering experience and experimental data, it is necessary to study the corresponding data mining methods and develop data mining tools. To solve this problem for typical electronic equipment, the following data mining methods are proposed, and then data mining tools are developed. Taking the cavity filter as an example, a mechanical model of the influence of manufacturing accuracy on electrical performance is established [1].

5.2.1 Data Modeling Method

In typical electronic equipment, whether it is to analyze its electrical performance or conduct system integration, the traditional method is usually used to convert the mechanical structure into a certain equivalent model and then use mature theories for analysis or comprehensive design. However, in the equivalent process, the influence of nonlinear factors or manufacturing errors is usually ignored, which makes the equivalent model have large difference with the actual one, resulting in poor reliability of electrical performance analysis results.

In response to this situation, a data mining method for modifying intermediate electrical parameters is proposed. The basic idea is to make full use of the existing empirical formulas or equivalent models and apply data mining techniques to these empirical formulas based on the data collected in the project, or the equivalent model is modified to obtain the mechanism model of the influence of mechanical structure factors on the electrical performance. Figure 5.2 shows the process of applying data mining methods to modify the intermediate electrical parameters to obtain the influence mechanism model.

In the design of electronic equipment, according to the preset electrical property indexes, through the existing methods, the desired intermediate electrical parameter M^0 and the corresponding structural dimensions $X^0 = \left[x_1^0, x_2^0, \ldots, x_n^0\right]^T$ in the circuit model can be obtained. However, in actual manufacturing, the processing errors, assembly errors, and surface roughness will inevitably be introduced, making the ideal structure size change $X = X^0 + \Delta X$, which will cause the actual

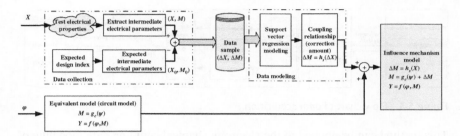

Figure 5.2 Flow chart of data mining method for correcting intermediate electrical parameters.

intermediate electrical parameters M^0 in the circuit model to change ΔM relative to the expected intermediate electrical parameters, so that the electrical performance of the system does not meet the expected electrical property indexes. Since the manufacturing accuracy will affect the intermediate electrical parameters in the circuit model, we may assume that there is the following mapping relationship between them as:

$$\Delta M = h_s(\Delta X) \tag{5.1}$$

The electrical performance of microwave devices in electronic equipment can usually be calculated using the existing circuit model. For the convenience of expression, it is assumed that the calculation formula for electrical performance is

$$M = g_c(\psi) \tag{5.2}$$

$$Y = f(\varphi, M) \tag{5.3}$$

in which M represents the intermediate electrical parameter in the circuit model, which is a function of the inductance, capacitance, excitation, or frequency in the equivalent circuit (represented by a vector ψ), and the electrical properties Y of microwave devices generally include insertion loss and return loss, etc. It is a function of intermediate electrical parameters M and other parameters φ (such as excitation and center frequency).

Using data mining technology to establish the correction value of the intermediate electrical parameters (5.1), and then using the intermediate electrical parameters M as a bridge to establish a mechanism model of the influence of mechanical structural factors on electrical performance, it can be expressed as

$$M = g_c(\psi) + h_s(\Delta X)$$

$$Y = f(\varphi, M) \tag{5.4}$$

The abovementioned method modifies the existing equivalent circuit model so that the modified model can contain more mechanical structure factors.

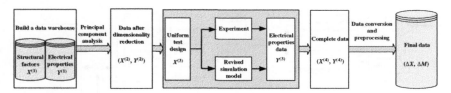

Figure 5.3 Flow chart of data acquisition.

This method can be applied to the electromechanical coupling data modeling of microwave devices or microelectromechanical systems.

5.2.2 Acquisition of Data Samples

The premise of establishing the expression of Eq. (5.1) is to obtain enough useful data samples. The acquisition process is shown in Figure 5.3.

5.2.2.1 Building the Initial Data Warehouse

The data warehouse is the basis for using data mining techniques to analyze the influence law of mechanical structure factors on electrical performance and establish its influence mechanism model. The data in the data warehouse can come from the product test data accumulated in the project, the test data of the test sample, or the prior knowledge in the project such as the simulation data generated by the simulation model.

In electronic equipment, a data warehouse usually consists of a database of mechanical structural factors and a database of corresponding electrical performance or intermediate electrical parameters. For the convenience of discussion, it may be assumed that the initial data sample set is

$$
\boldsymbol{X}^{(1)} \equiv
\begin{bmatrix}
x_{11}^{(1)} & x_{12}^{(1)} & \cdots & x_{1\bar{p}}^{(1)} \\
x_{21}^{(1)} & x_{22}^{(1)} & \cdots & x_{2\bar{p}}^{(1)} \\
\vdots & \vdots & \cdots & \vdots \\
x_{N_1 1}^{(1)} & x_{N_1 2}^{(1)} & \cdots & x_{N_1 \bar{p}}^{(1)}
\end{bmatrix}
\quad
\boldsymbol{Y}^{(1)} =
\begin{bmatrix}
y_{11}^{(1)} & y_{12}^{(1)} & \cdots & y_{1\bar{m}}^{(1)} \\
y_{21}^{(1)} & y_{22}^{(1)} & \cdots & y_{2\bar{m}}^{(1)} \\
\vdots & \vdots & \cdots & \vdots \\
y_{N_1 1}^{(1)} & y_{N_1 2}^{(1)} & \cdots & y_{N_1 \bar{m}}^{(1)}
\end{bmatrix}
\tag{5.5}
$$

where $\boldsymbol{X}^{(1)} \in R^{N_1 \times \bar{p}}$, $\boldsymbol{Y}^{(1)} \in R^{N_1 \times \bar{m}}$, and N_1 is the total number of initial samples, $\bar{p}$ is the total number of mechanical structure factors, and $\bar{m}$ is the total number of corresponding electrical performance indexes or intermediate electrical parameters.

5.2.2.2 Obtaining the Data Samples Needed for Modeling

In some special cases, the data in the original data warehouse can directly form the data samples needed for modeling, but in most cases, the original database needs a series of processing to obtain the data samples needed for modeling.

Principal Component Analysis In the original database, whether it is the structural parameter or electrical performance, there may be some nonindependent and unimportant factors. These so-called fake data are not only useless to find the influence mechanism but also may lead to no real influence relationship. Therefore, they must be excluded, and the way to achieve this goal is principal component analysis. Supposing that when the principal component analysis is passed, the original data sample is transformed into the following new data sample:

$$
X^{(2)} \equiv \begin{bmatrix} x_{11}^{(2)} & x_{12}^{(2)} & \cdots & x_{1\widetilde{p}}^{(2)} \\ x_{21}^{(2)} & x_{22}^{(2)} & \cdots & x_{2\widetilde{p}}^{(2)} \\ \vdots & \vdots & \cdots & \vdots \\ x_{N_1 1}^{(2)} & x_{N_1 2}^{(2)} & \cdots & x_{N_1\widetilde{p}}^{(2)} \end{bmatrix} \quad Y^{(2)} = \begin{bmatrix} y_{11}^{(2)} & y_{12}^{(2)} & \cdots & y_{1\widetilde{m}}^{(2)} \\ y_{21}^{(2)} & y_{22}^{(2)} & \cdots & y_{2\widetilde{m}}^{(2)} \\ \vdots & \vdots & \cdots & \vdots \\ y_{N_1 1}^{(2)} & y_{N_1 2}^{(2)} & \cdots & y_{N_1\widetilde{m}}^{(2)} \end{bmatrix} \tag{5.6}
$$

where $X^{(2)} \in R^{N_1 \times \widetilde{p}}$ and $Y^{(2)} \in R^{N_1 \times \widetilde{m}}$ are the new data sample sets obtained after principal component analysis; in general, $\widetilde{p} \le \overline{p}$, $\widetilde{m} \le \overline{m}$.

Uniform Design Although principal component analysis excludes relevant or unimportant data, there may be situations where the data distribution of a certain size is not good, such as structural parameters $x_{i\overline{p}}^{(2)} = (10, 80, 90, 100, 110, \ldots)$. The distance between 10 and 80 is obviously too far, and several data such as 20, 30, 40, 50, 60, and 70 need to be added. In general, to determine which data should be added, the uniform design method can be used. Supposing the sample of additional structural factors obtained in this way is N_2,

$$
X^{(3)} \equiv \begin{bmatrix} x_{11}^{(3)} & x_{12}^{(3)} & \cdots & x_{1\widetilde{p}}^{(3)} \\ x_{21}^{(3)} & x_{22}^{(3)} & \cdots & x_{2\widetilde{p}}^{(3)} \\ \vdots & \vdots & \cdots & \vdots \\ x_{N_2 1}^{(3)} & x_{N_2 2}^{(3)} & \cdots & x_{N_2\widetilde{p}}^{(3)} \end{bmatrix} \tag{5.7}
$$

The corresponding electrical properties samples are

$$
Y^{(3)} = \begin{bmatrix} y_{11}^{(3)} & y_{12}^{(3)} & \cdots & y_{1\widetilde{m}}^{(3)} \\ y_{21}^{(3)} & y_{22}^{(3)} & \cdots & y_{2\widetilde{m}}^{(3)} \\ \vdots & \vdots & \cdots & \vdots \\ y_{N_2 1}^{(3)} & y_{N_2 2}^{(3)} & \cdots & y_{N_2\widetilde{m}}^{(3)} \end{bmatrix}
$$

The abovementioned data samples N_2 can be obtained through two channels, one is to directly use the actual electronic equipment experimental test, and the other is to obtain the revised simulation model calculation. The following section describes how to use the simulation model to obtain these data.

Correction of Simulation Model According to formulas (5.1)–(5.4), the relationship between structural factors X and electrical properties Y can be simplified as

$$
Y = H(X, \zeta) \tag{5.8}
$$

where the parameter ζ is the function of ψ and φ in the original formula, and the operator H is the abbreviated form of the operator h_s, g_c, and f.

In Eq. (5.8), the influence of nonlinear structural factors (such as manufacturing error, assembly error, and surface roughness) is usually not considered, and when these factors are taken into account, the parameters ζ in the type of estimation of Eq. (5.8) (such as conductivity and magnetic permeability) will change. If the parameters ζ that consider the abovementioned nonlinear factors and satisfy the existing samples $(X^{(2)}, Y^{(2)})$ are found, then a modified simulation model will be obtained. A mathematical programming problem is specially constructed in the following form:

> Find: ζ
>
> Min: $\|Y - Y^{(2)}\|_2^2$
>
> s.t. $Y = H(X^{(2)}, \zeta)$
>
> $\zeta_l \leq \zeta \leq \zeta_u$ (5.9)

where ζ_u and ζ_l are the upper and lower limits of ζ, respectively.

Substituting the obtained ζ^* by solving the programming problem into Eq. (5.8) gives the following revised simulation model:

$$Y = H(X, \zeta^*) \tag{5.10}$$

According to the uniform design, using the revised simulation model (5.10), the additional data samples $(X^{(3)}, Y^{(3)})$ can be obtained. Then, combining the newly obtained sample $(X^{(3)}, Y^{(3)})$ with the sample $(X^{(2)}, Y^{(2)})$ to get the final data sample $(X^{(4)}, Y^{(4)})$,

$$X^{(4)} \equiv \begin{bmatrix} x_{11}^{(4)} & x_{12}^{(4)} & \cdots & x_{1\widetilde{p}}^{(4)} \\ x_{21}^{(4)} & x_{22}^{(4)} & \cdots & x_{2\widetilde{p}}^{(4)} \\ \vdots & \vdots & \cdots & \vdots \\ x_{N1}^{(4)} & x_{N2}^{(4)} & \cdots & x_{N\widetilde{p}}^{(4)} \end{bmatrix} \quad Y^{(4)} = \begin{bmatrix} y_{11}^{(4)} & y_{12}^{(4)} & \cdots & y_{1\widetilde{m}}^{(4)} \\ y_{21}^{(4)} & y_{22}^{(4)} & \cdots & y_{2\widetilde{m}}^{(4)} \\ \vdots & \vdots & \cdots & \vdots \\ y_{N1}^{(4)} & y_{N2}^{(4)} & \cdots & y_{N\widetilde{m}}^{(4)} \end{bmatrix} \tag{5.11}$$

where $N = N_1 + N_2$.

5.2.2.3 Data Conversion and Normalized Processing

What needs to be pointed out is that no matter $X^{(4)}$ or $Y^{(4)}$, there may be different magnitudes and dimensions, which is prone to ill-conditioned problems. To solve this, first, convert and normalize these data. Then, the processed data sample $X^{(5)}$ can be compared with X^0 to get ΔX. At the same time, the corresponding intermediate electrical parameter $M^{(5)}$ can be obtained from the processed data sample $Y^{(5)}$ and the basic knowledge of electrical properties, which can then be compared with M^0 to get ΔM. Finally, the data samples $(\Delta X, \Delta M)$ needed to establish the relationship (5.1) can be obtained.

5.2.3 Multicore Regression Method for Data Mining

After obtaining the samples $(\Delta X, \Delta M)$, the data mining procedure based on the linear combination of conventional basis functions is required. In order to improve the accuracy of data mining modeling, it may be possible to assume (5.1) as the following multicore function form:

$$h_s(\Delta x) = \sum_{r=1}^{l} \sum_{i=1}^{N} \alpha_{ri} k_r(\Delta x, \Delta x_i) + b \tag{5.12}$$

where l is the total number of kernels, and $k_r(\Delta x, \Delta x_i)$ is the rth kernel function. The introduction of the kernel function solves the regression problem of nonlinear and high-dimensional data and avoids the difficulty of direct searching for nonlinear mapping. The commonly used kernel functions include Gaussian kernel, polynomial kernel, wavelet kernel, and perceptron kernel. The selection of kernel functions depends on the characteristics of the problem. In general, polynomial kernel is more suitable for the problem of gentle change, and Gaussian kernel and wavelet kernel are more suitable for the problem of steep change. α_{ri} ($r = 1, 2, \ldots, l$; $i = 1, 2, \ldots, N$) is the coefficient that needs to be found; b is the constant that needs to be found.

Obviously, this is a multicore fitting function. The reason for using multicore is that the curve or surface to be approximated or fitted is complex in practical problems. For example, there are both flat areas and steep areas, and there are mutations. It is difficult to complete the task with a single kernel function.

In order to obtain the unknown variables, when the input and output samples are known, the unknown variables can be obtained by solving the following programming problem:

$$\text{Min: } \sum_{r=1}^{l} c_r \|\alpha_r\|_1 + 2C \sum_{i=1}^{N} L(\Delta M_i - f_i(x))$$

$$\text{s.t.} - \sum_{r=1}^{l} c_r \sum_{j=1}^{N} \alpha_{rj} k_r(\Delta x_i, \Delta x_j) - b + \Delta M_i \le \varepsilon + \xi_i$$

$$\sum_{r=1}^{l} c_r \sum_{j=1}^{N} \alpha_{rj} k_r(\Delta x_i, \Delta x_j) + b - \Delta M_i \le \varepsilon + \xi_i$$

$$\xi_i \ge 0 \quad (\forall i = 1, 2, \cdots, N). \tag{5.13}$$

where the first term in the objective function represents the generalization ability of the regression function, and the second term represents the fitting ability of the regression function to the training samples. By selecting appropriate parameters, the accuracy of the regression function can be increased, and the generalization ability of the regression function can be improved. In the data modeling of the cavity filter, the parameter C generally takes a value between 10 and 200. The vector

$\boldsymbol{\alpha}_r = [\alpha_{r1}, \alpha_{r2}, \ldots, \alpha_{rN}]^T$ is the model complexity corresponding to the *rth* kernel function, which satisfies

$$\|\boldsymbol{\alpha}_r\|_1 = \sum_{i=1}^{N} |\alpha_{ri}| \tag{5.14}$$

If the number of elements α_{ri} equal to zero in $\boldsymbol{\alpha}_r$ is greater, the regression model obtained is sparser. c_r is used to punish the nonzero variable α_{ri}. The value of c_r is related to the kernel parameter of the corresponding kernel function, and c_r can be taken as the reciprocal of the kernel parameter. For example, if the Gaussian kernel $k(\Delta x, \Delta x_i) = \exp\left[-\frac{\|\Delta x - \Delta x_i\|^2}{2\sigma^2}\right]$ is used, $c_r = \frac{1}{\sigma}$ is used, and the value of σ should be determined according to the specific problem. In the following specific calculations, the value range is between 0 and 1. The ε-insensitive cost loss function is defined as

$$L(\Delta M_i - f_i(\boldsymbol{x})) = \begin{cases} 0, & \text{if} |\Delta M - f_i(\boldsymbol{x})| \leq \varepsilon \\ |\Delta M - f_i(\boldsymbol{x})| - \varepsilon, & \text{otherwise} \end{cases} \tag{5.15}$$

where ε is the allowable overtolerance, and its value depends on the specific problem. In the modeling of microwave devices, ε can take a value in $[0.001, 0.01]$.

When solving (5.13), because of the absolute value of the variable, we can introduce

$$\alpha_{ri} = \alpha_{ri}^+ - \alpha_{ri}^- \quad |\alpha_{ri}| = \alpha_{ri}^+ + \alpha_{ri}^- \tag{5.16}$$

where $\alpha_{ri}^+, \alpha_{ri}^- \geq 0$.

If α_i is known, only one of the above two equations is satisfied, and the two variables cannot be greater than zero at the same time, that is $\alpha_{ri}^+ \cdot \alpha_{ri}^- = 0$ needs to be satisfied.

Substituting Eq. (5.16) into Eq. (5.13) leads to the following nonlinear programming problem:

Find: $\alpha_{ri}^+, \alpha_{ri}^-, \xi_i, b$

Min: $\displaystyle\sum_{r=1}^{l} c_r \sum_{j=1}^{N} \left(\alpha_{rj}^+ + \alpha_{rj}^-\right) + 2C \sum_{i=1}^{N} \xi_i$

s.t. $-\displaystyle\sum_{r=1}^{l}\sum_{j=1}^{N} \left(\alpha_{rj}^+ - \alpha_{rj}^-\right) k_r(\Delta \boldsymbol{x}_i, \Delta \boldsymbol{x}_j) - b + \Delta M_i \leq \varepsilon + \xi_i$

$\displaystyle\sum_{r=1}^{l}\sum_{j=1}^{N} \left(\alpha_{rj}^+ - \alpha_{rj}^-\right) k_r(\Delta \boldsymbol{x}_i, \Delta \boldsymbol{x}_j) + b - \Delta M_i \leq \varepsilon + \xi_i$

$\alpha_{ri}^+ \geq 0, \alpha_{ri}^- \geq 0, \xi_i \geq 0. \quad (r = 1, 2, \cdots l, i = 1, 2, \cdots N) \tag{5.17}$

By solving this problem, the required design variables can be obtained.

5.2.4 Application of Data Mining

To verify the correctness and effectiveness of the above method, it is especially applied to the modeling of the influence mechanism of the cavity filter, and some satisfactory results are obtained.

Cavity filters are widely used in various electronic devices. Its main function is to transmit signals within a specified frequency range as much as possible while suppressing noise outside the specified frequency range. Cavity filters usually consist of multiple resonant cavities, adjusting bolts, and input and output coupling ring. For a filter including n resonant cavities, if the size of the mechanical structure changes by ΔX, the coupling matrix will be changed to ΔM, and the electrical properties of the filter, such as insertion loss and return loss, will also be changed correspondingly.

Assuming that some data sample set $(\Delta X, \Delta M)$ has been obtained, then according to the data modeling method given in Section 5.2.1, the mechanism model of the influence of manufacturing accuracy on the electrical performance of the filter can be established.

$$\Delta M = \widehat{h}_s(\Delta X) \tag{5.18}$$

$$S_{21}(f, \Delta X) = -2j\sqrt{R_1 R_2}\left[\left(\frac{f_0}{BW}\left(\frac{f}{f_0} - \frac{f_0}{f}\right)I - jR + M_0 + \Delta M\right)^{-1}\right]_{n1}$$

$$S_{11}(f, \Delta X) = 1 + 2j\sqrt{R_1}\left[\left(\frac{f_0}{BW}\left(\frac{f}{f_0} - \frac{f_0}{f}\right)I - jR + M_0 + \Delta M\right)^{-1}\right]_{11} \tag{5.19}$$

where $\widehat{h}_s(\Delta X)$ is the model obtained using the support vector regression algorithm, $S_{21}(f_i, \Delta X)$ and $S_{11}(f_i, \Delta X)$ are the insertion loss and return loss corresponding to the f_ith frequency point, respectively, and I, BW, f_0, f, and n, respectively, represent the unit array, expected bandwidth, center frequency, operating frequency, and filter series. In the matrix R, except for the main diagonal $R_{11} = R_1$, $R_{nn} = R_2$, and $R_{ii} = \frac{f_0}{BW \cdot Q}(1 < i < n)$, all other elements are zero, where R_1 and R_2 are the source impedance and the load impedance, Q is the no-load quality factor, and $\widetilde{M}$ is the designed coupling matrix.

The influence mechanism established above can be applied to the debugging of the cavity filter, that is the optimal insertion depth of the adjustable bolt of the filter can be obtained through the following optimization problem:

Find: ΔX_m

Min: $\displaystyle\sum_{f_i=sf}^{ef}\left[\left(S_{21}(f_i, \Delta X) - S_{21}^m(f_i)\right)^2 + \left(S_{11}(f_i, \Delta X) - S_{11}^m(f_i)\right)^2\right]$

s.t. $\Delta X_m^L \leq \Delta X_m \leq \Delta X_m^U \quad (m = 1, 2)$ \hfill (5.20) where ΔX_m, ΔX_m^U, and ΔX_m^L represent the actual, upper, and lower limits of the change of the filter adjustment bolt, respectively, and sf ef are the starting and stopping frequency values, respectively. If $m = 1$, $S_{21}^m(f_i)$ and $S_{11}^m(f_i)$ represent the insertion loss and return loss of the current filter measured by the vector network analyzer, respectively. On the other hand, if $m = 2$, both $S_{21}^m(f_i)$ and $S_{11}^m(f_i)$ are the predefined filter debugging objectives.

According to Eq. (5.20), the auxiliary debugging process of the constructed cavity filter is as shown in Figure 5.4. When debugging, it is necessary to pre-debug the assembled filter first and then obtain the adjustment amount of each bolt according to Eq. (5.20). The specific steps are as follows:

(1) Use a vector network analyzer to measure the electrical properties of the filter after pre-debugging, that is $S_{21}^1(f_i)$ and $S_{11}^1(f_i)$, and then solve the optimization problem (5.20) in $r = 1$ and get the depth $L_1 = L_0 + \Delta X_1$ of each adjustment bolt, where L_0 denotes the depth benchmark of the filter-adjusting bolt, which is determined by the synthesis or experiment of the filter.

(2) According to the predetermined debugging targets $S_{21}^2(f_i)$ and $S_{11}^2(f_i)$, the optimization problem (5.20) in $r = 2$ is solved, and the screwing depth $L_2 = L_0 + \Delta X_2$ of each bolt can be obtained.

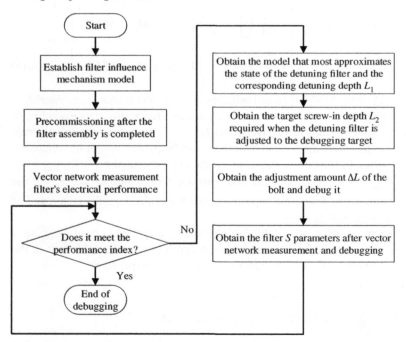

Figure 5.4 Auxiliary debugging process of cavity filter.

(3) From L_1 and L_2, the adjustment amount of each bolt of the filter can be obtained as $\Delta D = L_2 - L_1 = \Delta X_2 - \Delta X_1$.

The influence mechanism of the mechanical structural factors on the electrical performance of the filter and the auxiliary debugging method of the filter based on the above influence mechanism are obtained. The following is a test of a waveguide cavity filter to verify its correctness and effectiveness.

Figure 5.5 shows a waveguide filter. Its center frequency, bandwidth, and return loss are 397.7 MHz, 8.31 MHz, and 20 dB, respectively. It is necessary to solve the problem of low filter debugging efficiency. By the previous data mining method, according to the debugging experience data in the project, apply the support vector regression algorithm to establish the influence mechanism between the adjustable bolt of the filter and its electrical performance and then use the mechanism model to obtain the screwing depth of seven adjusting bolts of the filter (Figure 5.5).

The electrical properties of the filter include insertion loss S_{21} and return loss S_{11}. According to the debugging experience accumulated in the project, 45 data samples were collected and sorted out, and the final data sample set can be obtained after principal component analysis. Considering the space limitations, Table 5.1 only lists 45 data samples corresponding to the operating frequency of 397.5 MHz.

To ensure the effectiveness of the influence mechanism in the application of filter debugging, another five sets of waveguide filters are used to proof the correctness of the mechanism model. Table 5.2 shows the comparison result of the model prediction and the measured data for 397.5 MHz, and Figure 5.6 shows the prediction, measured, and its error curve in the frequency range of 385–410 MHz.

It can be seen from Table 5.2 that when the operating frequency is 397.5 MHz, the maximum relative error between the results calculated by influence mechanism and the measured results does not exceed 11%. At the same time, it can also

(a)   (b)

Figure 5.5 The structure of the waveguide filter and the physical photo of the auxiliary debugging system. (a) Structure of waveguide filter. (b) Auxiliary debugging system.

Table 5.1 45 data samples at 397.5 MHz.

Sample number	Bolt 1 length (mm)	Bolt 2 length (mm)	Bolt 3 length (mm)	Bolt 4 length (mm)	Bolt 5 length (mm)	Bolt 6 length (mm)	Bolt 7 length (mm)	S_{21} (dB)	S_{11} (dB)
1	71.5	70.5	72	75.8	68.4	68	70.6	−1.7472	−25.4637
2	71.3	70.1	72.3	76.2	68.6	68.4	70.6	−2.0428	−12.2880
3	71.5	70.3	72.1	75.6	68.6	68.8	70.9	−2.0544	−12.5194
4	71.3	70.9	72.1	76.4	68.3	68.6	70.6	−1.5983	−21.6266
5	71.5	70.7	72	76.2	68.5	68.6	70.9	−1.9260	−14.7746
6	71.2	70.7	72	76.4	68.6	68	70.7	−1.8372	−14.0791
7	71.1	70.3	72.1	76.2	68.3	68	70.8	−2.2334	−10.4528
8	71.3	70.5	72.3	75.6	68.5	68.2	71	−1.9797	−14.8086
9	71.4	70.9	72.1	76.2	68.3	68.8	70.9	−2.0494	−17.3222
10	71.2	70.9	72.1	76	68.7	68.8	70.8	−1.9535	−17.4945
11	71.4	70.3	71.9	76	68.7	68.6	70.6	−3.5386	−5.9830
12	71.1	70.5	72.3	76.4	68.7	68.6	70.9	−2.4330	−10.7969
13	71.2	70.9	72.3	76	68.4	68	70.9	−2.6005	−9.4028
14	71.1	70.9	72	75.8	68.6	68.4	71	−1.9796	−18.9334
15	71.4	70.7	72.2	75.8	68.3	68.6	71	−2.1345	−14.7125
16	71.4	70.9	71.9	75.6	68.4	68.4	70.8	−1.7916	−18.3826
17	71.2	70.1	71.9	75.8	68.7	68.2	70.9	−2.1905	−12.2068
18	71.4	70.3	72.2	76.4	68.4	68.2	70.7	−1.9040	−15.5999
19	71.3	70.7	71.9	76.2	68.3	68.2	70.9	−1.9600	−13.9610
20	71.2	70.3	72.2	76.2	68.4	68.8	71	−2.1277	−13.0212
21	71.3	70.5	72.2	75.6	68.7	68	70.8	−1.7335	−18.6864
22	71.3	70.1	72	76.4	68.5	68.8	70.8	−1.7531	−19.0827
23	71.2	70.1	72	75.6	68.3	68.6	70.7	−1.7554	−18.9117
24	71.4	70.1	72.1	76	68.6	68	71	−1.8927	−14.1994
25	71.3	70.5	72.1	76	68.5	68.4	70.8	−1.7877	−21.4746
26	71.1	70.7	72.1	75.6	68.5	68.2	70.6	−1.7791	−23.8705
27	71.1	70.5	71.9	76	68.4	68.8	70.7	−1.8334	−35.1381
28	71.5	70.9	72.2	76.2	68.7	68.2	70.7	−1.9221	−22.2985
29	71.5	70.1	72.3	76	68.3	68.4	70.8	−1.9708	−14.8026
30	71.5	70.5	71.9	76.4	68.5	68.4	71	−2.0913	−12.5485
31	71.4	70.7	72.3	75.8	68.6	68.8	70.7	−1.8210	−22.3673
32	71.3	70.9	72.3	76	68.7	68.2	70.9	−1.66720	−16.8466

Table 5.1 (Continued)

Sample number	Bolt 1 length (mm)	Bolt 2 length (mm)	Bolt 3 length (mm)	Bolt 4 length (mm)	Bolt 5 length (mm)	Bolt 6 length (mm)	Bolt 7 length (mm)	S_{21} (dB)	S_{11} (dB)
33	71.3	70.7	72.3	75.6	68.3	68.4	70.7	−2.1818	−12.0808
34	71.1	70.9	72	75.6	68.4	68.6	70.8	−1.8594	−18.6674
35	71.4	70.1	72	76.4	68.6	68.4	70.9	−2.3643	−10.6014
36	71.5	70.3	72.3	76.2	68.4	68	70.8	−1.9523	−15.5265
37	71.5	70.3	72.2	75.6	68.5	68.6	71	−1.7330	−23.0083
38	71.1	70.1	72.3	76	68.5	68.6	70.6	−1.8606	−15.5728
39	71.2	70.1	72.1	75.8	68.3	68.4	70.9	−1.6367	−19.2949
40	71.2	70.7	72.2	76.4	68.6	68.6	70.8	−2.0970	−13.9376
41	71.5	70.7	71.9	76.2	68.7	68.4	70.7	−2.9475	−7.6196
42	71.5	70.5	72	76	68.3	68.2	70.6	−1.8775	−15.0002
43	71.2	70.5	72	76	68.7	68.8	71	−2.1180	−16.3618
44	71.1	70.5	72.2	76.4	68.4	68.2	71	−2.7381	−8.4166
45	71.3	70.1	72.1	75.6	68.7	68	70.7	−1.9962	−13.1963

Table 5.2 Measured and predicted data of five sets of filters (397.5 MHz).

Verified sample number	S_{21} (dB)			S_{11} (dB)		
	Measured	Prediction	Relative error (%)	Measured	Prediction	Relative error (%)
1	−1.7330	−1.7383	0.3018	−23.008	−24.6029	6.9306
2	−1.9774	−1.9803	0.1491	−15.987	−16.2621	1.7209
3	−1.7400	−1.7376	0.1351	−18.8826	−19.1581	1.4590
4	−1.6711	−1.7026	1.8901	−19.5939	−17.5198	10.5854
5	−1.7880	−1.7877	0.0157	−21.3527	−20.8983	2.1278

be noticed, from Figure 5.6, that in the entire operating frequency band the error is relatively small. It shows that the established influence mechanism meets the needs of the project requirement and can be applied to the auxiliary debugging of the filter.

Applying the mentioned verified influence mechanism, a debugging method of the filter is constructed. Figure 5.7 shows the bolt adjustment amount by

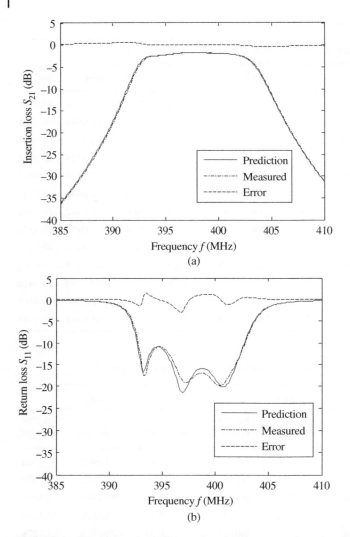

Figure 5.6 Comparison between predicted and measured results of influence mechanism (385–410 MHz). (a) Comparison of insertion loss. (b) Comparison of return loss.

applying this method. The adjustment amount indicates the direction and size of each adjustment bolt screwing in. Figure 5.8 shows the detuning state of the filter before commissioning, the preset commissioning target, the measurement results after commissioning, and the electrical performance error curve. These curves show that the electrical performance of the filter after commissioning is

Figure 5.7 Adjustment amount of bolt (397.5 MHz).

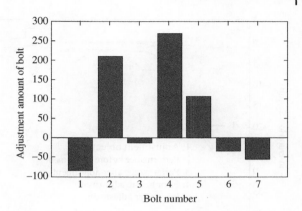

better than the electrical performance before commissioning, to get closer to the debugging target.

It can be seen that the adjustment amount obtained according to the debugging method can assist the engineering and technical personnel to complete the debugging of the filter, and it only needs to be done once. This method can overcome the disadvantages of the need for experienced debugging personnel, thoughtlessness debugging, and long debugging cycles in the project and significantly improve the debugging efficiency. With the continuous accumulation of debugging experience data in the project, it should be noted that the accuracy of the debugging method could be increased accordingly.

5.3 ISFP of Reflector Antennas

5.3.1 Data Collection and Mining

To obtain the influence mechanism of the reflector antenna structure factors on the electrical performance, a large amount of engineering data and experiment data were collected and sorted first, and a data warehouse was established in a prescribed format. The collected engineering test data covers four kinds of apertures of reflector antennas of 9, 12, 13, and 16 m. They are working at C and Ku frequency bands. In detail, there are 6 antennas with 9-m and C-band, 6 antennas with 9-m and Ku-band, 20 antennas with 12-m and C-band antennas, 6 antennas with 13-m and Ku-band antennas, 4 antennas with 16-m and C-band antennas, and 5 antennas with 16-m and C/Ku dual-band. The collected engineering test data includes measured electrical performance (including gain, first sidelobe level, and efficiency) and corresponding mechanical structural factors (including precision of single panel, precision of reflector assembly, the supporting position

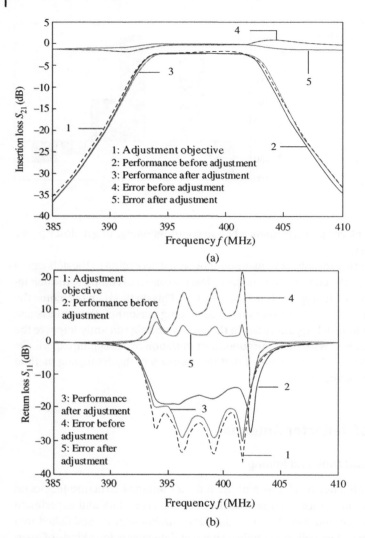

Figure 5.8 Comparison of electrical properties of detuning filter before and after adjustment (385–410 MHz).(a) Insertion loss. (b) Return loss.

of subreflector, the projection area of the supporting leg of subreflector, and the cross-sectional geometric dimensions of the subreflector bracket).

In addition, the 3.7 m Ku-band reflector antenna was used to design an experimental system and conduct actual tests on structural factors and electrical performance. The basic method is to divide the reflector into a number of rings and zones, add gaskets of different thicknesses to produce different deformations, and then measure the errors of each shape (including single-panel accuracy, the thickness of the gasket, and the position of the gasket) and the corresponding far-field pattern to reflect the influence of structural factors on electrical performance. To include the possible error distribution in the project as much as possible, so that the electrical performance test results could truly reflect the influence of structural factors on the electrical performance, the deformation of the circular ring domain, the sector domain, and the symmetrical and asymmetrical reflection surface is considered in the experiment. Through practical tests, 109 sets of data were obtained, which laid the foundation for the establishment and verification of the influence mechanism.

5.3.2 The Establishment of an Analysis Model of the Influence Mechanism

The theory of the reflector antenna tells us that the aperture field distribution and the far-field distribution are the relationships of Fourier transform pairs (Figure 5.9), and the far-field pattern of the antenna can be described mathematically as

$$E(\theta, \phi) = \iint_A E_0(\rho', \phi') e^{jk\rho' \sin\theta \cos(\phi-\phi')} \rho' d\rho' d\phi' \tag{5.21}$$

The meaning of each parameter in the formula is shown in Figure 3.2 in Chapter 3.

If there is a phase error in the aperture plane, the integration of the electric field on the aperture plane needs to know the functional expression of the phase error, which is difficult to obtain to some extent. For the sake of this, the aperture plane

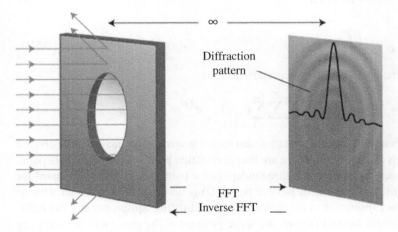

∞

Diffraction pattern

FFT
Inverse FFT

Figure 5.9 The relationship between aperture field and far field.

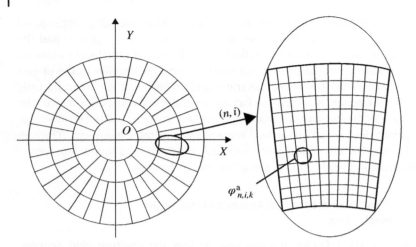

Figure 5.10 Schematic diagram of the aperture plane corresponding to the reflective panel.

needs to be discretized. As shown in Figure 5.10, the aperture plane is divided into grids according to the division of the reflective panel. It may be assumed to set a total of N circles; each circle has K_n blocks, each block has a grid unit, and each grid center point corresponds to the aperture plane phase. The phase error is $\varphi_{n,i,k}^a$.

The far-field electric field is the superposition of the Fourier transform of the surface field at each small unit, that is

$$E(\theta, \phi) = \sum_{n=1}^{N} \sum_{i=1}^{K_n} \sum_{k=1}^{P_n} E_{n,i,k} e^{j\varphi_{n,i,k}^a} \tag{5.22}$$

where

$$E_{n,i,k} = \iint_{A(n,i,k)} E_0(\rho', \phi') e^{jk\rho' \sin\theta \cos(\phi - \phi')} \rho' d\rho' d\phi' \tag{5.23}$$

Then, the far-field radiation pattern is

$$P = EE^* = \sum_{n=1}^{N} \sum_{m=1}^{N} \sum_{i=1}^{K_n} \sum_{j=1}^{K_m} \sum_{k=1}^{P_n} \sum_{l=1}^{P_m} E_{n,i,k} E_{m,j,l}^* e^{j\left(\varphi_{n,i,k}^a - \varphi_{m,j,l}^a\right)} \tag{5.24}$$

The reflective surface of the large- and medium-sized antennas is mostly divided into panels (Figure 5.11). There are four connection points between a single panel and the back frame of which three independent points (A, B, and C) are used for adjustment, and the fourth passive point (D) is used for reinforcement. Due to the accuracy limitation of the measuring and adjusting equipment and the influence of various human factors, the space position of the panel will deviate from its ideal state. These assembly errors will cause the phase error of the antenna

Figure 5.11 Schematic diagram of the spatial position of the panel deviation.

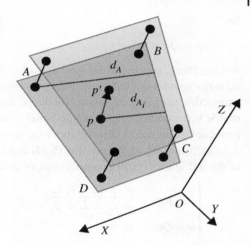

aperture plane field and affect the antenna's far-field pattern. Because the surface deformation during panel debugging and installation is small and can be ignored, the deviation motion during panel assembly can be regarded as rigid body movement, and there is a certain relationship between the deviation of the panel installation point and the phase error of the aperture plane.

As mentioned before, the reflecting surface is composed of $\sum$ panels, and the k panel has P_k nodes. According to the assumption of the rigid body panel, if the normal deviations of the three mounting points A, B, and C of the kth panel are a_A, a_B, and a_C, respectively, the phase error of the aperture plane corresponding to the ith node caused by this deviation is

$$\varphi_i^k = B_i^k \cdot a^k \tag{5.25}$$

where B_i^k is the conversion matrix between the deviation of the kth panel and the ith phase error, and $a^k = [a_A, a_B, a_C]^T$ is the deviation of the three adjustment points of the kth panel.

If there are P_k nodes on a single panel, the phase error is $\varphi^k = [\varphi_1{}^k, \varphi_2{}^k, \ldots,$ $\varphi_{P_k}^k]^T$, and the conversion matrix is $B^k = \left[B_1{}^k, B_2{}^k, \ldots, B_{P_k}^k \right]^T$. So, there is a phase error $\varphi = [\varphi^1, \varphi^2, \ldots, \varphi^\Sigma]^T$ for all panels of the entire reflecting surface, the adjustment vector $a = [a^1, a^2, \ldots, a^\Sigma]^T$, and the conversion matrix

$$Q = \begin{bmatrix} B^1 & 0 & \ldots & 0 \\ 0 & B^2 & \ldots & 0 \\ \ldots & \ldots & \ldots & \ldots \\ 0 & 0 & \ldots & B^\Sigma \end{bmatrix}$$

So

$$\varphi = Qa \tag{5.26}$$

Among them, $\varphi = \{\varphi_i\}$ is the phase error of the aperture plane corresponding to all surface nodes caused by the panel deviation; Q is the overall conversion matrix connecting the panel deviation and the surface phase error, and its diagonal block matrix is the corresponding single-panel conversion matrix B^k; and $a = \{a_j\}$ is the panel deviation vector defined above the mounting point number j.

Substituting Eq. (5.26) into Eq. (5.24) yields the following analysis relationship of the influence of panel assembly errors on the antenna radiation pattern [2–6]:

$$\begin{cases} P = EE^* = \sum_{n=1}^{N} \sum_{m=1}^{N} \sum_{i=1}^{K_n} \sum_{j=1}^{K_m} \sum_{k=1}^{P_n} \sum_{l=1}^{P_m} E_{n,i,k} E_{m,j,l}^* e^{j\left(\varphi_{n,i,k}^a - \varphi_{m,j,l}^a\right)} \\ \varphi = Qa \end{cases} \tag{5.27}$$

5.3.3 Experiment

Here, the 3.7 m Ku-band rotating paraboloid antenna mentioned in Section 3.2.6 is still used as a verification case with different loading methods. The purpose here is to make the reflector deviate. Therefore, three fan-shaped panels located in the lower left area of the reflector (Figure 5.12) are selected to add gaskets, and the gaskets with thickness of 3 mm are added at 13 bolts. According to the established influence mechanism analysis model, the equal area method is adopted to divide the reflecting surface into triangular meshes (Figure 5.13). The side length of the grid triangle is $\lambda/4$, and the whole reflecting surface is

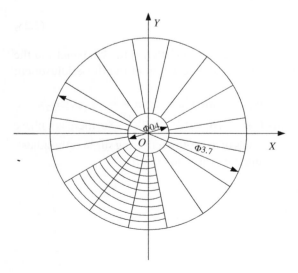

Figure 5.12 Schematic diagram of antenna reflector panel distribution.

Figure 5.13 Schematic diagram of electromagnetic grid division.

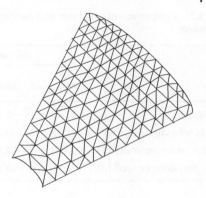

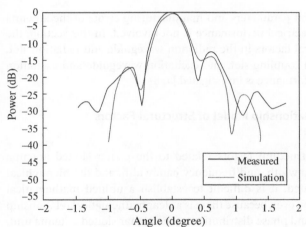

Figure 5.14 Comparison of simulated and measured values of the far-field pattern.

divided into $21904 \times 12 = 262\,848$ elements. The far-field pattern of the antenna is calculated by Eq. (5.33).

Comparing the simulated values of the far-field pattern after the deviation of the selected three panels with the actual test results, as shown in Figure 5.14 and Table 5.3, it can be seen that the calculated value of the simulation and the measured value are in good agreement at the main lobe, and the left and right first sidelobes also basically coincide.

5.4 ISFP of Planar Slotted Waveguide Array Antennas

The influence of the deformation of the planar slotted antenna array on its electrical performance has been studied from the perspective of field coupling. The

Table 5.3 Comparison of simulated and measured values of antenna electrical performance parameters.

	Normal 0 mm			Normal 3 mm		
	Measured	Simulation	Errors (%)	Measured	Simulation	Errors (%)
Gain (dB)	52.23	52.84	1.17	52.00	52.21	0.4
3 dB beamwidth (degrees)	0.380	0.390	2.63	0.49	0.48	2.04
Left first sidelobe level (dB)	−14.54	−14.33	1.44	−17.8	−17.3	2.8
Right first sidelobe level (dB)	−14.18	−14.33	1.06	−12.53	−13.85	10.5

influence of the structural parameters and manufacturing errors of the antenna cavity and slot on the electrical performance is not involved. In this section, the influence of the structural factors in the radiation waveguide and radiation slot, coupling waveguide and coupling slot, and excitation waveguide and excitation slot on the electrical performance is investigated [7, 8].

5.4.1 Hierarchical Relationship Model of Structural Factors and Electrical Properties

Since there are many structural factors related to the planar slotted antenna gain, sidelobe level, beamwidth, and frequency bandwidth, and the hierarchical relationship is complicated, it is difficult to establish a unified mathematical model between them. However, because there is a clear mathematical relationship between the amplitude and phase distribution of the planar slotted antenna unit, and the antenna electrical performance indexes and the number of structural factors corresponding to the amplitude and phase indexes of the unit are relatively small, the hierarchical relationship is relatively simple. Therefore, the feasible method is to divide the relationship between the structural factors and the electrical properties of the planar slotted antenna into two levels. One is the relationship between electrical performance indicators and unit amplitude and phase errors, and the other is the relationship between the structural factors of functional components (referring to radiation, coupling, and excitation waveguides and slots) and the unit amplitude and phase errors, which is shown in Figure 5.15.

By establishing a two-level relationship model, the mechanism of the influence of the structural factors of the planar slotted array antenna on the electrical performance can be focused on the mechanism between the structural factors in the antenna functional components and the unit amplitude and phase distribution, so as to obtain the influence relationship of the structural factors of the antenna functional components and the amplitude and phase relative indicators of the unit.

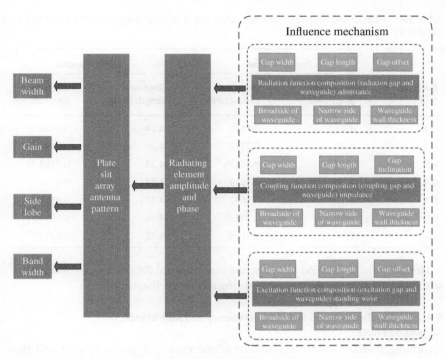

Figure 5.15 Two-level relationship model of planar slotted array antenna.

5.4.2 Influence of Structural Factors on the Amplitude Phase of a Unit in a Radiated Functional Component

Radiation function component refers to radiation waveguide and the corresponding radiation gap. In the radiation function components, the structural factors such as the shape, position, size of the slot, and the thickness of the waveguide have varying degrees of influence on the amplitude and phase error of the unit. The specific analysis is as follows:

5.4.2.1 Influence of Slot Deviation on Conductance and Resonance Length

The effects of radiation gap bias on conductance and resonant length are shown in Table 5.4.

Taking the circular head gap as an example, it can be seen by fitting that the resonant length A and conductance B are the polynomial functions of bias C.

$$l = 0.3788x^4 - 0.7828x^3 + 0.6250x^2 + 0.1973x + 4.1912 \tag{5.28}$$

$$\varphi = -38.8333x^5 + 135.3598x^4 - 184.2742x^3 + 122.5746x^2 - 38.7937x + 4.7867 \tag{5.29}$$

Table 5.4 Relationship between slot deviation and resonance length (round and square head slot).

Deviation (mm)	Rounded head		Square headed	
	Resonant length (mm)	Electrical conductance	Resonant length (mm)	Electrical conductance
0.4	4.33	0.1549	4.14	0.1580
0.5	4.37	0.2472	4.20	0.2447
0.6	4.42	0.3532	4.23	0.3493
0.7	4.45	0.4649	4.27	0.4570
0.8	4.51	0.5657	4.31	0.5897
0.9	4.55	0.7023	4.36	0.6994
1.0	4.61	0.8196	4.41	0.8213

This formula is obtained by fitting the experimental data of a single waveguide with a single gap, and the following fitting formula is the same.

5.4.2.2 The Relationship Between Frequency and Admittance, Amplitude Phase

Assuming that the deviation distance of the radiation slot is 1.0 mm and the resonance length is 4.61 mm, the relationship among conductance, susceptance, amplitude phase, and frequency can be obtained (Figure 5.16).

5.4.2.3 Influence of Waveguide Wall Thickness on Admittance, Amplitude Phase

Assuming that the deviation distance is 1.0 mm and the resonance length is 4.61 mm, the relationship among admittance, amplitude phase, and waveguide wall thickness is as shown in Figure 5.17.

5.4.2.4 Influence of Slot Width on Admittance, Amplitude Phase

Assuming that the deviation distance is 1.0 mm and the resonance length is 4.61 mm, the influence of the slot width on admittance and amplitude is as shown in Figure 5.18.

5.4.2.5 Influence of Slot Length on Amplitude and Phase

The amplitude and phase variations caused by the slot length at different deviations are in the range of 0.3–1.9% in amplitude and 1.47–4.01 degrees in phase, corresponding to an error increment of 0.01 mm in the slot length.

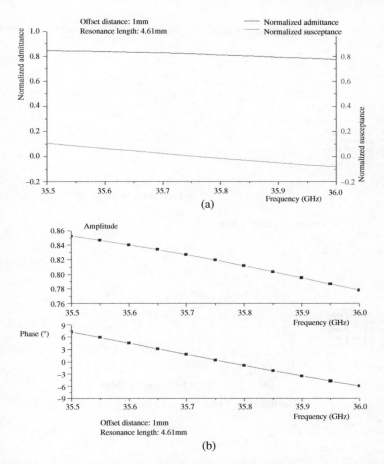

Figure 5.16 The relationship among normalized conductance, susceptance, amplitude phase, and frequency.

5.4.3 Influence of Structural Factors on the Amplitude Phase of a Unit in a Coupling Functional Component

The coupling function component refers to the coupling waveguide and the corresponding coupling gap. The structural factors such as the direction (inclination angle), size, and waveguide wall thickness of the gap have different degrees of influence on the amplitude and phase of the unit, which are described as follows:

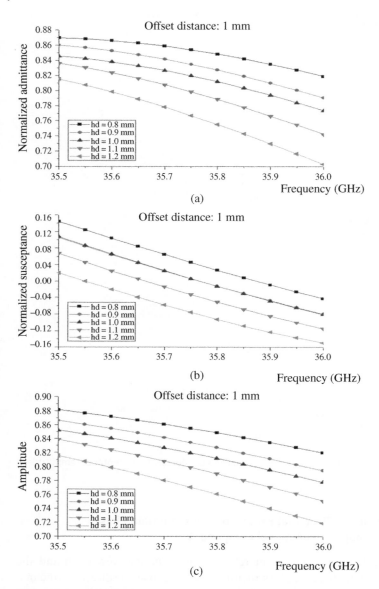

Figure 5.17 The influence of waveguide wall thickness on admittance, susceptance, amplitude, and phase. (a) Influence of waveguide wall thickness on normalized admittance. (b) Influence of waveguide wall thickness on normalized susceptance. (c) Influence of waveguide wall thickness on amplitude. (d) Influence of waveguide wall thickness on phase.

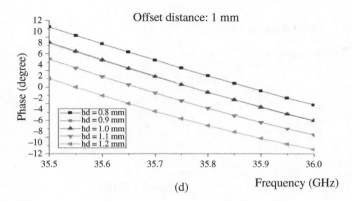

Figure 5.17 (*Continued*)

Table 5.5 The relationship between inclination angle with resonance length and normalized resonance resistance.

θ (°)	10	11	12	13	14	15
Resonant length (mm)	4.54	4.53	4.53	4.53	4.535	4.54
Normalized resistance	0.103	0.127	0.151	0.177	0.203	0.233
θ (°)	16	17	18	19	20	21
Resonant length (mm)	4.54	4.55	4.55	4.555	4.56	4.565
Normalized resistance	0.263	0.295	0.33	0.364	0.401	0.438
θ (°)	22	23	24	25	26	27
Resonant length (mm)	4.57	4.58	4.58	4.59	4.6	4.6
Normalized resistance	0.478	0.518	0.561	0.604	0.646	0.691
θ (°)	28	29	30			
Resonant length (mm)	4.61	4.62	4.62			
Normalized resistance	0.734	0.781	0.831			

5.4.3.1 Influence of the Inclination Angle of the Slot on the Resonance Length and Resonance Resistance

Firstly, the influence of slot inclination angle (10–30°) on resonant length and normalized resonant resistance is studied. The experimental results of the self-made sample are shown in Table 5.5.

The fitting formula of the relationship between resonant resistance and inclination angle can be obtained from the experimental data in Table 5.5.

$$r = 0.0009\theta^2 + 0.0036\theta - 0.0227, \theta \in (0, 30) \tag{5.30}$$

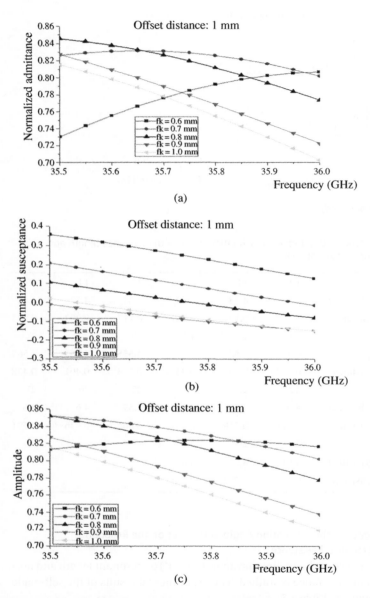

Figure 5.18 The influence of slot width on admittance, susceptance, amplitude phase. (a) Influence of slot width on normalized admittance. (b) Influence of slot width on normalized susceptance. (c) Influence of slot width on amplitude. (d) Influence of slot width on phase.

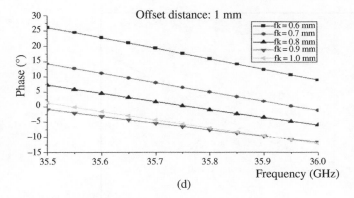

(d)

Figure 5.18 (*Continued*)

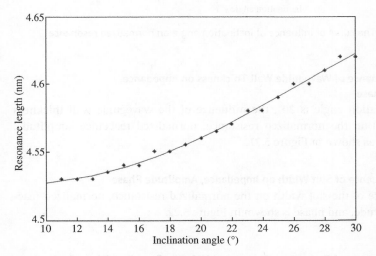

Figure 5.19 The curve of influence of inclination angle on resonance length.

The corresponding influence curve can be obtained (Figure 5.19, 5.20).

5.4.3.2 Influence of Inclination Angle and Slot Length on Amplitude and Phase

Corresponding to the error increment of 0.01 mm in gap length will cause the phase variations in the range of 0.9–2.538° and the amplitude variations in the range of 0.049–1.669%. Corresponding to the increment of inclination angle error per one degree, the amplitude variations in the range of 6–11.5% and the phase variations in the range of 0.6–2.2° are caused.

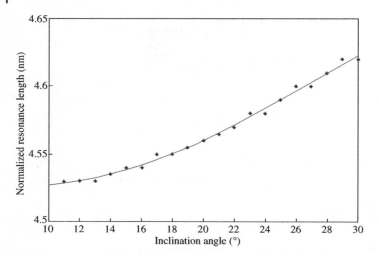

Figure 5.20 The curve of influence of inclination angle on normalized resonance resistance.

5.4.3.3 Influence of Waveguide Wall Thickness on Impedance, Amplitude Phase

If the inclination angle is 30°, the influence of the waveguide wall thickness (0.8–1.2 mm) on the normalized resistance, normalized reactance, amplitude, and phase is as shown in Figure 5.21.

5.4.3.4 Influence of Slot Width on Impedance, Amplitude Phase

The influence of the slot width on the normalized resistance, normalized reactance, amplitude, and phase is shown in Figure 5.22.

5.4.4 Influence of Structural Factors on Voltage Standing Wave Ratio in the Excitation Functional Components

The excitation functional component refers to the excitation waveguide and its corresponding excitation slot, and its main electrical performance index is the voltage standing wave ratio, which is 1 in the ideal condition, and the actual value is related to structural factors such as slot deviation, slot width, slot length, and waveguide wall thickness. The influence of these structural factors on the slot standing wave is given by simulation.

The excitation slot model based on HFSS simulation is shown in Figure 5.23, and the model parameters, frequency, wavelength, and other parameters are shown in Table 5.6.

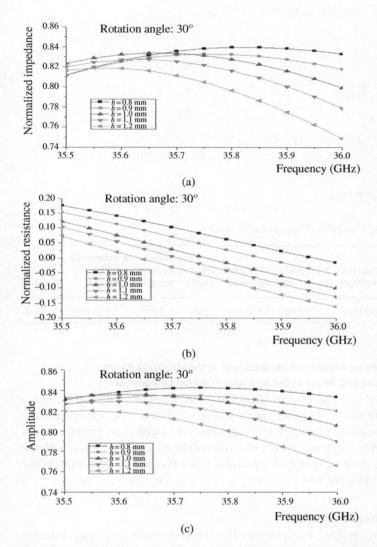

Figure 5.21 Influence of waveguide wall thickness on impedance, resistance, amplitude phase. (a) Influence of waveguide wall thickness on impedance. (b) Influence of waveguide wall thickness on resistance. (c) Influence of waveguide wall thickness on amplitude. (d) Influence of waveguide wall thickness on phase.

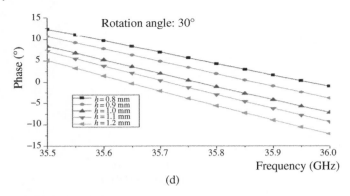

Figure 5.21 (*Continued*)

Table 5.6 The parameters of the excitation model.

Frequency	Atmosphere wavelength	Waveguide wavelength	Waveguide thickness	Crevices widths	Waveguide narrower margin	Waveguide wide margin	Deviation
35.75 GHz	8.392 mm	12.992 mm	0.5 mm	1 mm	1.5 mm	5.496 mm	1.05 mm

5.4.4.1 Weighting Analysis of the Influence of the structural Factors on the Amplitude and Phase in the Incentive Function Component

The orthogonal test was used to select the tolerance of all structural factors as ±0.01 mm, and the level of each factor was selected as 3. The orthogonal table L18 (37) was selected to arrange the test. Then the electrical performance value (standing wave) of each set of test parameters of the excitation joint is calculated, and the influence weight of each structural factor is obtained. The contribution rate is shown in Figure 5.24.

5.4.4.2 Results and Discussion

From the above analysis, it can be seen that when the same structural size tolerance is ±0.01 mm, the main structural factor that significantly affects the voltage standing wave ratio performance of the excitation slot is mainly the deviation (about 65%), followed by the broadside and narrow side of the waveguide (respectively, accounting for about 10%). In comparison, other structural factors have little effect on the performance of the excitation slot voltage standing wave ratio. Therefore, in practical engineering, the size tolerances of waveguide wide side, narrow side, and bias of excitation joint should be strictly limited.

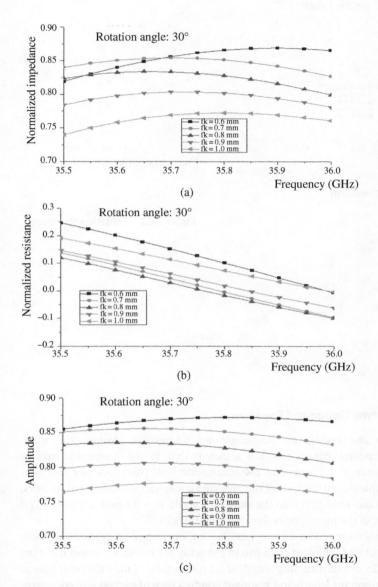

Figure 5.22 Influence of slot width on impedance, resistance, amplitude phase.
(a) Influence of slot width on impedance. (b) Influence of slot width on resistance.
(c) Influence of slot width on amplitude. (d) Influence of slot width on phase.

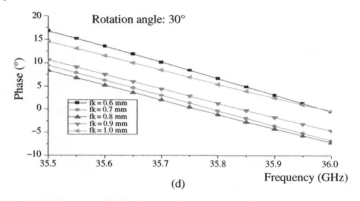

(d)

Figure 5.22 *(Continued)*

Figure 5.23 Excitation slot model.

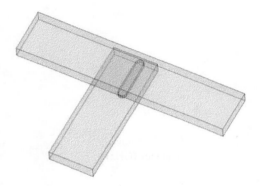

5.4.5 Prototype Design and Experiment

According to the results of the single-slot admittance design, the same waveguide structure parameters are selected, that is the waveguide size is $a \times b = 5.496\,\text{mm} \times 1.5\,\text{mm}$, and the waveguide wall thickness is $t = 1\,\text{mm}$. According to the weight analysis of the previous single-slot structural parameters, the slot width has little effect, so the same slot width $w = 0.8\,\text{mm}$ is chosen. The design process of the prototype is shown in Figure 5.25.

The moments method is used to establish a uniform small array model of 5×5, that is the offset and length of each gap in the array are the same. Select different biases from 0.25 to 1.05 mm, step length of 0.1 mm, a total of nine different biases; calculate the resonant length and resonant conductance of each slot in the corresponding array; and then get the normalized admittance curve and fitting formula of each slot. The admittance curve of the first gap is shown in Figure 5.26.

The fitting formula is

$$I(\delta) = -0.1861\delta^3 + 0.347\delta^2 + 0.0302\delta + 4.1469 \tag{5.31}$$

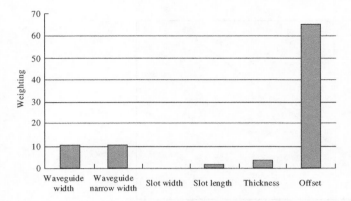

Figure 5.24 Weighting of the influence factor of the structure parameter of the excitation slot.

Figure 5.25 Design flow of test sample.

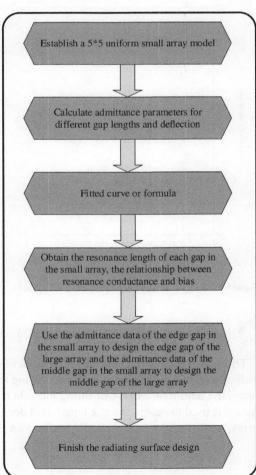

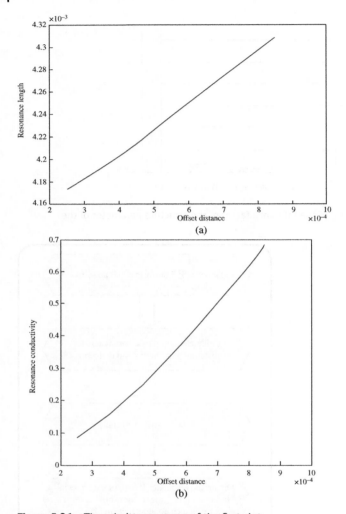

Figure 5.26 The admittance curve of the first slot.

$$g(\delta) = -0.2306\delta^3 + 0.9083\delta^2 + 0.2327\delta - 0.0286 \tag{5.32}$$

The admittance data of the above slots can be used to design the slots of the radiation surface in the large array. According to the aperture distribution of the array, the admittance curve or fitting formula of the edge slot in the small array model is used to design the slot length and deviation of the edge slot in the large array, while the slot in the middle position on the radiation surface of the large array is designed by the admittance curve or fitting formula of the middle slot in the middle position on the small array model.

5.5 ISFP of Microwave Feeder and Filters

Section 5.2 discussed a data mining approach based on the relationship between the influence of structural factors on electrical performance based on massive data and applied it to the electrically tunable duplex and the waveguide filters. This section discusses the influence of structural factors on the electrical performance of microwave feeders and filters through an analytical, deductive approach [9–14]. The structural factors and electrical properties involved are shown in Table 5.7.

5.5.1 Hierarchical Relationship Model of the Influence of Structural Factors on the Resonant Cavity Filters

Through theoretical analysis and engineering experience, combined with expert opinions, the corresponding relationship between the main structural factors and

Table 5.7 The research content of the influence mechanism of mechanical structure factors on the resonant cavity filter.

Study Subjects	Structural parameters	Manufacturing accuracy	Electrical performance parameters	Research content
Resonant cavity (physics) filters	Structural form Structural dimensions Tuning method material Surface treatment	Dimensional accuracy Shape accuracy Position accuracy Surface quality Plating quality Tuning accuracy flaw	Center frequency Bandwidth fluctuations Insertion loss, standing waves bandstop suppression In-band time delay, the power capacity	Mutual influence relationships, including theoretical models and relationships, empirical formulas, measured (including simulation) relationship diagrams, tables, engineering experience, qualitative relationships and description of potential laws, etc.

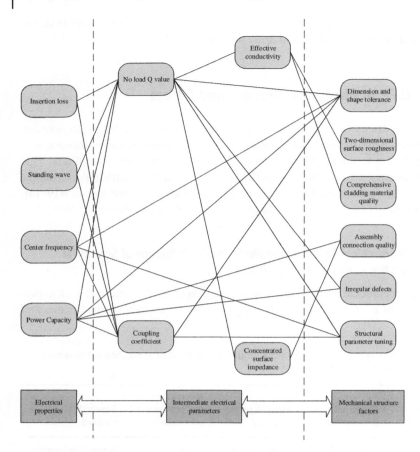

Figure 5.27 Hierarchical relationship model between structural factors and electrical performance.

electrical performance of the resonant cavity filter and the importance classification table can be obtained. Based on this idea, the hierarchical relationship model between the structural factors and the electrical properties is constructed by intermediate variables (Figure 5.27).

Based on the refined relationship model between mechanical structural factors and electrical properties, the following technical approaches are obtained:

(1) To establish the analysis model between the intermediate electrical parameters and electrical performance of the filter.

The relationship model between intermediate electrical parameters (equivalent surface conductivity, concentrated surface resistance, the no-load Q value of single resonator, input/output coupling coefficient, and coupling coefficient between resonators) and filter characteristics (passband insertion loss and passband bandwidth) is established.

(2) To build up the relationship between the filter mechanical structure factors and the main intermediate electrical parameters.

   (a) Study the relationship model and influence of surface roughness, coating quality, and equivalent surface conductivity among mechanical structural factors.

   (b) Study the relationship model and influence of different structural forms, assembly connection quality, and concentrated surface resistance among mechanical structural factors.

   The relationship model and influence mechanism of the cavity size, the inner surface quality of the cavity, cavity forming method, tuning method, and other factors in the mechanical structure factors on the no-load Q value of the resonant cavity are studied, which are carried out by theoretical research, simulation, and prototype production.

   (c) Study the relationship model and influence mechanism of input/output and interstage coupling form, size, position, and input/output coupling coefficient and interstage coupling coefficient among mechanical structural factors and conduct theoretical research, simulation, and production of prototypes.

   (d) The influence of mechanical structure factors on tuning is evaluated mainly by simulation, and the influence of machining accuracy on the resonance frequency under the tuning mode of the plate capacitor is analyzed. The influence of the endplate diameter and endplate spacing of flat capacitors on the resonance frequency is mainly investigated.

   (e) In view of the influence of mechanical structure factors on the power capacity of microwave filters, the power capacity changes of resonators with different structural forms and machining defects are studied.

5.5.2 Influence of Structural Factors on the No-load Q Value of the Resonant Cavity

The hierarchical relationship between structure factors and no-load Q value for resonator filters is shown in Figure 5.28.

The analysis of the influence of the main structural factors on the unloaded Q value is as follows:

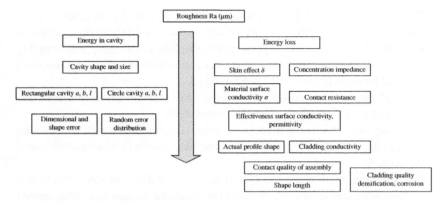

Figure 5.28 Hierarchical relationship of structural factors on the no-load Q value.

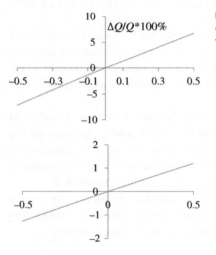

Figure 5.29 Influence of the deviation of cavity geometry b and l on the no-load Q value.

5.5.2.1 Influence of Geometric Shape, Size, and Position Deviation on the No-load Q Value

According to the hierarchical relationship between the structural factors affecting the no-load Q value, the curve can be obtained by theoretical analysis and simulation (Figures 5.29–5.32).

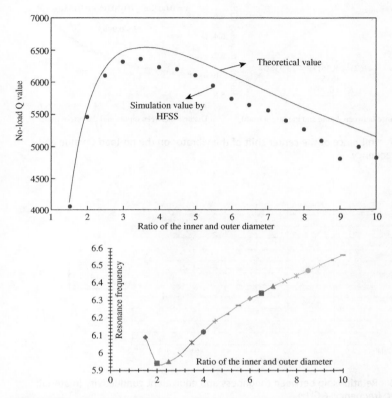

Figure 5.30 The influence of the ratio of inner to outer diameter of the coaxial cavity on no-load Q value and resonance frequency.

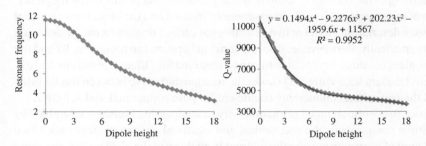

Figure 5.31 Influence of vibrator height on resonance frequency and no-load Q value.

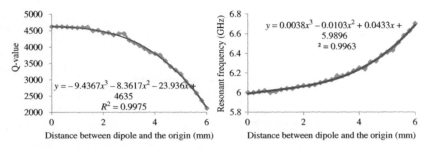

Figure 5.32 Influence of the center shift of the vibrator on the no-load Q value and resonant frequency.

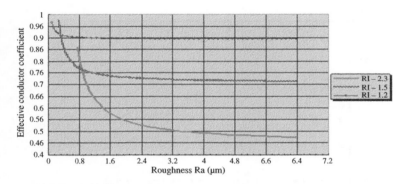

Figure 5.33 Relationship between roughness and equivalent conductivity (material: pure copper; frequency: 6 GHz).

5.5.2.2 Relationship Between Surface Roughness and Equivalent Conductivity

According to the concept of radio frequency conductivity profile of the rough surface of microwave devices, the measurement method and the evaluation technical index, a description model of two-dimensional surface roughness can be deducted mathematically. Furthermore, the relationship between the roughness Ra and the equivalent conductivity coefficient can be obtained for different materials and different frequencies. Figure 5.33 shows the relationship curve between the Ra value and the equivalent conductivity coefficient for pure copper material at 6 GHz.

Since the common metal resonant cavity in engineering is always assembled by multiple components, the mechanical and electrical characteristics of each face and joint of the cavity are greatly different from those of the ideal model, and there is obvious inconsistency, and the influence relationship is much more complex. Therefore, the influence of the roughness of the coaxial cavity at different positions

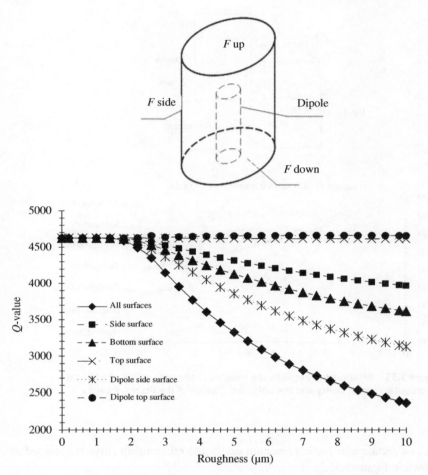

Figure 5.34 Relationship between the roughness of each surface of a typical circular coaxial cavity and the no-load Q value of the resonant cavity.

on the no-load Q is explored, and the relationship curve between the roughness of the circular coaxial resonator and the no-load Q value of the resonator is obtained (Figure 5.34).

From the above analysis, it can be seen that the roughness of the various faces of the resonant cavity affects the Q value in the following order: the bottom face of the cavity (-21.79%), the sides of the cavity (-14.1%), and the top face of the cavity (-0.17%).

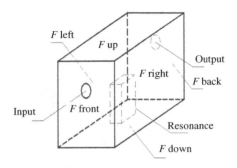

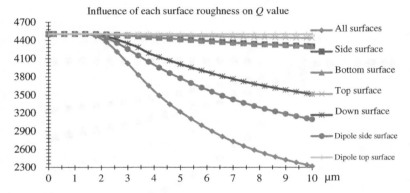

Figure 5.35 Relationship between the roughness of each surface of a typical rectangular coaxial cavity and the unloaded Q value of the resonant cavity.

For a rectangular coaxial resonant cavity, the relationship curve is obtained as shown in Figure 5.35.

5.5.2.3 Relationship Between Coating Quality and Equivalent Conductivity

The main influencing factors of coating quality on surface conductivity are coating thickness, coating purity, coating density, and coating surface roughness. Supposing the equivalent conductivity of the coating is

$$\sigma_c = \alpha_1 \alpha_2 \alpha_3 \alpha_4 \sigma \tag{5.33}$$

where α_1, α_2, α_3, and α_4 represent the coefficients of the four main factors mentioned above on the equivalent conductivity of the coating, respectively.

Considering theory analysis and engineering practice let us know that the comprehensive effect of coating quality on equivalent conductivity is quite obvious. In general case, the equivalent conductivity of coating is about 60–80% of the conductivity of ideal coating materials.

5.5.2.4 Influence of Coaxial Cavity Assembly Connection Quality on No-load Q Value

Machining (such as milling, welding, and splicing) or assembly methods of the coaxial cavity may lead to additional contact resistance between the components of the coaxial cavity. The additional contact resistance and the original surface resistance of the components of the coaxial cavity cause the loss of electromagnetic energy in the cavity. Therefore, in order to know the influence of assembly connection quality on the unloaded Q value, the independent surface of the contact area is added purposely (Figure 5.36).

It can be seen from Figure 5.36 that the quality of the assembly connection has the most obvious influence on the electrical performance at the contact point between the resonator and the bottom plate. Correspondingly, its concentrated contact impedance is of the greatest weight on electrical performance. So, during the designing and processing, the surface current concentration should be as smooth as possible to avoid machined defects such as burrs and slip marks. In addition, the direction of the machined texture and the direction of surface current flow should be aligned as closely as possible. This will significantly improve the electrical performance of devices or equipment.

5.5.3 Influence of Structural Factors on the Coupling Coefficient

In the resonant cavity filter, structural factors such as the size of the coupling hole, the position of the diaphragm, and the length of the resonator rod have obvious effect on the coupling coefficient, which are described as follows:

5.5.3.1 Influence of Coupling Hole Structure Factors on the Coupling Coefficient

The relationship curve of the length and width of the coupling hole on the coupling coefficient is shown in Figure 5.37. The relationship between the wall thickness of the coupling hole and the ratio of the inner and outer diameters of the resonant cavity on the coupling coefficient is shown in Figure 5.38.

5.5.3.2 Analysis of the Influence of the Position and Size of the Coupling Diaphragm and the Length of the Resonant Rod on the Coupling Coefficient

Establishing the interstage coupling simulation model for the coaxial cavity filter will give the relationship between the position and size of the coupling diaphragm and the length of the resonant rod on the coupling coefficient, as shown in Figure 5.39.

5.5.4 Influence of Tuning Screw on Resonance Frequency and Coupling Coefficient

5.5.4.1 Effect of Screw-in Depth on Resonant Frequency

The variable trend of the resonant frequency of the coaxial cavity with the depth of the tuning screw is shown in Figure 5.40.

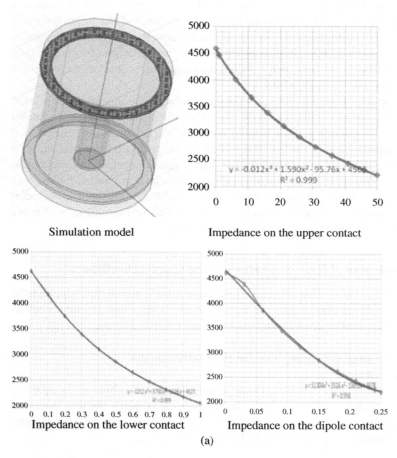

Simulation model

Impedance on the upper contact

$y = -0.012x^3 + 1.590x^2 - 95.76x + 4566$
$R^2 = 0.999$

Impedance on the lower contact

Impedance on the dipole contact

(a)

Figure 5.36 Influence of coaxial cavity assembly connection quality on no-load Q value. (a) Circular cavity. (b) Rectangular cavity.

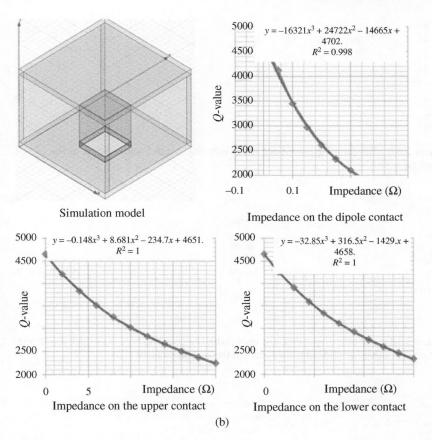

Simulation model

Impedance on the dipole contact

Impedance on the upper contact

Impedance on the lower contact

(b)

Figure 5.36 (*Continued*)

5.5.4.2 Relationship of the Influence of the Tuning Screw on the Coupling Coefficient

Suppose there is energy coupling between two resonant cavities with the same resonant frequency, the relationship of the influence of the coaxial cavity filter tuning screw on the coupling coefficient is as shown in Figure 5.41.

From the above analysis results, the coupling coefficient will gradually increase with the screwing length of the tuning screw. This tuning law can be employed to change the relative bandwidth for the adjustment of the filter.

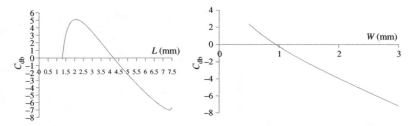

Figure 5.37 The coupling coefficient varies with the length *L* and width *W* of the coupling hole.

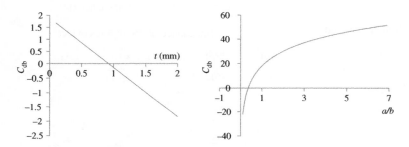

Figure 5.38 The coupling coefficient varies with the coupling hole wall thickness *T* and the ratio of the inner and outer diameters of the resonant cavity.

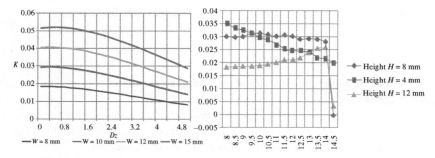

Figure 5.39 Relationship between the position and size of the coupling diaphragm and the length of the resonant rod on the coupling coefficient.

5.5.5 Influence of Structural Factors on the Power Capacity of Microwave Filters

To sum up, the factors affecting the power capacity of the filter can be roughly classified into three categories. One is related to the physical structure of the single

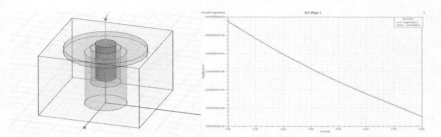

Figure 5.40 Relationship between the depth of the tuning screw and the resonance frequency.

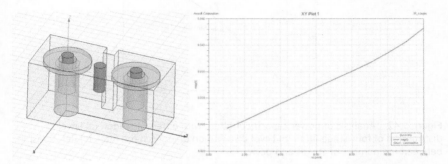

Figure 5.41 Influence of coaxial cavity filter tuning screws on the coupling coefficient.

cavity of the filter, the other is related to the characteristics of the filter itself, and the third is related to the environment in which the filter is located. For this reason, the influence of the number of stages of filter and relative bandwidth on the power capacity is studied separately, and more details are shown in Figures 5.42 and 5.43.

Comparing the results of the above analysis, it can be seen that the first and last cavities are always subjected to the smallest power, and both are less than the input power. As the number of stages increases, the maximum power that the intermediate cavity can endure gradually increases, 40.1 dBm for stage 4, 41.2 dBm for stage 5, 42.2 dBm for stage 6, 42.9 dBm for stage 7, and 43.2 dBm for stage 8, in the passband.

It can be seen that the first and the last cavities always bear the smallest power and are less than the input power. As the number of stages increases, the maximum power applied to the intermediate cavities increases in descending order, with 40.1, 41.2, 42.2, 42.9, and 43.2 dBm for stages 4, 5, 6, 7, and 8 in the passband, respectively.

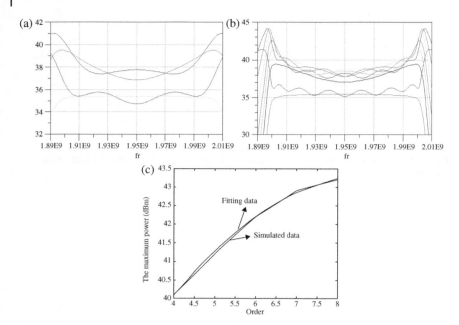

Figure 5.42 Analysis of power capacity of filters with different stages. (a) Level distribution of four-stage filter. (b) Level distribution of eight-stage filter. (c) Relationship between the number of stages and power.

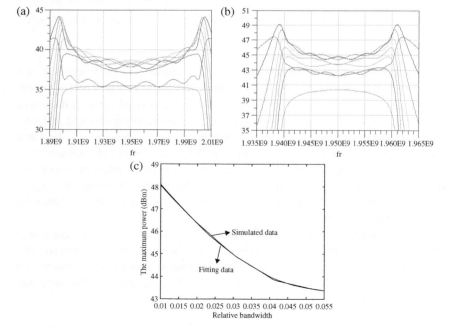

Figure 5.43 Simulation results of power capacity with the same number of stages and different bandwidths. (a) 100 MHz bandwidth level distribution. (b) 20 MHz bandwidth level distribution. (c) Relationship between bandwidth and power capacity.

Figure 5.44 Photos of some typical samples.

The above results also tell us that the last resonant cavity still has the least power. As the working bandwidth decreases, the power received by the last cavity will gradually increase, and the maximum power of the filter in the entire passband will gradually increase.

5.5.6 Prototype Production and Experiment

A large number of prototypes (Figure 5.44 shows some of the test samples) need to be made and tested in the process of research and verification of the influence mechanism. Based on the test results, an electromechanical coupling model of the filter prototype was established, and the influence of the roughness of the test prototype, the quality of the coating on the effective conductivity of the radio frequency, and the quality of the assembly connection on the concentrated contact impedance were analyzed to obtain the results of the analysis of the electromechanical coupling model of the filter prototype.

Table 5.8 shows the comparison between the analysis results of the filter sample influence mechanism and the measurement results. The difference between the analyzed value of the main electrical performances (insertion loss, standing wave, and passband width) and the measured value is located within 9–14%.

5.6 ISFP of Radar-Servo Mechanism

With the increasing demands on system performance such as radar beam pointing accuracy, the traditional method of separating structure and control design is

Table 5.8 Results of data analysis of filter test samples.

Sample number	Material	Structural form	Conductivity (theoretical S/m)	Insertion loss at f0 of ideal model	Insertion loss at f0 of pure metal model	Insertion loss at f0 of electro-mechanical coupling model	Measured insertion loss at f0	Error
142	Stainless aluminum		2.5 e7	−0.023	−0.086	−0.29	−0.32	9.4%
146	Gold-plated brass		4.55 e7	−0.023	−0.069	−0.27	−0.34	21%
135	Silver-plated brass		6.25 e7	−0.023	−0.067	−0.31	−0.34	8.8%
134	brass		1.5 e7	−0.023	−0.089	−0.49	−0.57	14%

becoming increasingly difficult to achieve, and it is imperative to integrate structure and control design into a unified framework in order to pursue optimal overall performance [15]. To realize the integrated design of structure and control, it is necessary to understand the mechanism of mechanical structural factors on the performance of the servo system and the influence of structural factors such as friction, clearance, inertia distribution, and structural support on the tracking performance of the servo system.

According to the characteristics of radar antenna-servo system, a method of integrating mass (inertia), friction, clearance, and other structural factors into the mass matrix, damping matrix, stiffness matrix, and excitation vector is proposed, and then the multibody dynamics equation considering the above nonlinear structural factors is derived, that is the analysis model of the influence mechanism of mechanical structural factors on the servo system. In addition, a servo experimental platform was developed to verify the correctness of the model.

5.6.1 Influence of Clearance on the Performance of the Servo System

5.6.1.1 Influence of Gear Meshing Clearance
Gear transmission is a transmission form widely used in radar antenna-servo systems [16]. To ensure the normal operation, a certain lateral clearance must be left between the tooth surfaces of two gears that mesh with each other, as shown in Figure 5.45. The presence of tooth backlash will not only cause transmission error but also cause oscillation or shock to reduce tracking performance. The method of introducing the tooth backlash into the system model is described as follows [17, 18]:

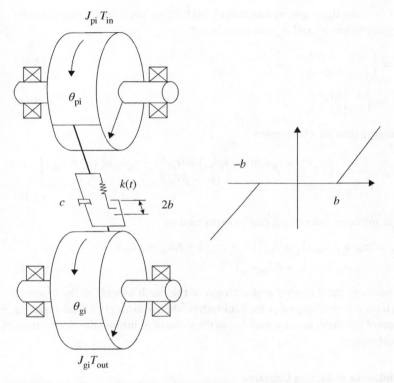

Figure 5.45 Schematic diagram of meshing gear with clearance.

Let k_{mi} and c_{mi} be the joggling stiffness and damping of the ith pair of meshing gears, δ_i the rotational displacement difference between the driving wheel and the driven wheel, namely, $\delta_i = r_{pi}\theta_i - r_{gi}\theta_i$, where θ_{pi} and θ_{gi} are the rotation angles of the driving gear and the driven gear, and $r_{pi}\ r_{gi}$ the radiuses of the driving gear and the driven gear, respectively.

If the meshing clearance is considered, the meshing force can be expressed as [19]

$$f(\delta_i, \dot{\delta}_i) = k_{mi} \begin{cases} (\delta_i - b_i)(1 + \beta_i \dot{\delta}_i) & , & \delta_i \geq b_i \\ 0 & , & -b_i < \delta_i < b_i \\ (\delta_i + b_i)(1 + \beta_i \dot{\delta}_i) & , & \delta_i \leq -b_i \end{cases} \quad (5.34)$$

where b_i is the tooth gap of the $b_i^* = \begin{cases} -b_i, & \delta \geq b_i \\ b_i, & \delta \leq -b_i \end{cases}$ th pair of meshing gears, and

the impact factor $n_{mi} = \begin{cases} 1 & |\delta_i| > b_i \\ 0 & |\delta_i| \leq b_i \end{cases}$ is generally taken as 0–0.2.

In order to make the segment function (5.34) become continuous function, the following parameters b_i^* and n_{mi} are introduced:

$$b_i^* = \begin{cases} -b_i, & \delta \geq b_i \\ b_i, & \delta \leq -b_i \end{cases} \tag{5.35}$$

$$n_{mi} = \begin{cases} 1 & |\delta_i| > b_i \\ 0 & |\delta_i| \leq b_i \end{cases} \tag{5.36}$$

And bringing into Eq. (5.34) gives

$$f(\delta_i, \dot{\delta}_i) = n_{mi} \left[\begin{array}{c} (r_{pi}\theta_{pi} - r_{gi}\theta_{gi})(1 + \beta_i\dot{\delta}_{mi}) + (r_{pi}\dot{\theta}_{pi} - r_{gi}\dot{\theta}_{gi})\beta_i \left(b_i^* + \delta_{mi} \right) \\ + \left(b_i^* - \beta_i\delta_{mi}\dot{\delta}_{mi} \right) \end{array} \right] \tag{5.37}$$

Thus, the meshing force of F_m can be expressed as

$$F_{mi} = \left(c_{mi} + \beta_i k_{mi} \left(b_i^* + \delta_{mi} \right) \right) \dot{\delta}_i + n_{mi}(1 + \beta_i\dot{\delta}_{mi})k_{mi}R_{pi}\delta_i$$
$$+ n_{mi}k_{mi} \left(b_i^* - \beta_i\dot{\delta}_{mi}\delta_{mi} \right) \tag{5.38}$$

As can be seen, the damping and stiffness of the mesh as well as the external excitation term will be changed as the backlash is taken into account. The damping and stiffness of the mesh are also variable as the value of b_i^* jumps with the rotation of the wheel system.

5.6.1.2 Influence of Bearing Clearance

In addition to the tooth backlash, bearing clearance also has an important influence on servo performance. If the bearing clearance is too large, it will not only reduce the resonance frequency of the system but also increase the rotation error of the shafting, thereby reducing the speed and accuracy of the system. On the contrary, if the bearing clearance is too small, it will increase the friction torque, which will not only increase the wear but also easily cause low-speed crawling and reduce the low-speed stability of the system. The method of introducing the clearance in the ball bearing into the system model is discussed below [20–22].

It is supposed that there is a clearance b_1 between the ball and the inner and outer rings, as shown in Figure 5.46. Assuming that the angle between the jth ball and the horizontal direction is ϕ_j, the radial displacement ξ_j between the inner and outer rings can be expressed as [23]

$$\xi_j = (x_s - x_p)\cos\phi_j + (y_s - y_p)\sin\phi_j - b_1 \tag{5.39}$$

where x_s, y_s, x_p, and y_p represent the displacement of the inner and outer rings of the bearing in the x and y directions, respectively.

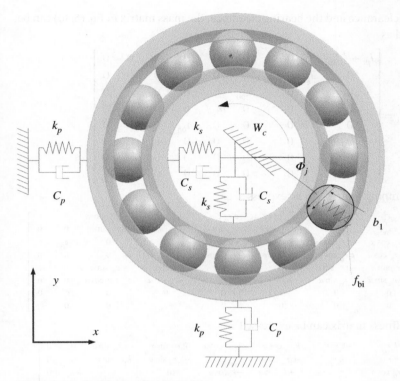

Figure 5.46 Bearing nonlinear model.

The coupling equation between the bearing and the gearing can be established by loading the full ball pressure in the horizontal and vertical directions as an excitation term into the gear rotation equation,

$$M_i \ddot{X} + C_i \dot{X} + K_i X = P_i \tag{5.40}$$

where the state variable is $X = \begin{bmatrix} \theta_{pi} & \theta_{gi} & x_{pi} & x_{gi} & y_{pi} & y_{gi} & x_s & y_s \end{bmatrix}^T$.

Supposing the mass and damping of the inner ring of the bearing are m_{bp} and c_{bp}, respectively, the total mass of the corresponding shaft and gear is m_s, and the contact stiffness and damping of the inner ring of the bearing and the shaft are k_s and c_s, respectively, and the meshing angle of the gear is α. After introducing the meshing clearance and the bearing clearance, the mass matrix in Eq. (5.40) can be expressed as

$$
M_i = \begin{bmatrix}
J_{pi} + \frac{1}{3}J_{spi} & 0 & 0 & 0 & 0 & 0 & 0 & 0 \\
0 & J_{gi} + \frac{1}{3}J_{sgi} & 0 & 0 & 0 & 0 & 0 & 0 \\
0 & 0 & m_{bp} & 0 & 0 & 0 & 0 & 0 \\
0 & 0 & 0 & m_{bg} & 0 & 0 & 0 & 0 \\
0 & 0 & 0 & 0 & m_{bp} & 0 & 0 & 0 \\
0 & 0 & 0 & 0 & 0 & m_{bp} & 0 & 0 \\
0 & 0 & 0 & 0 & 0 & 0 & m_s & 0 \\
0 & 0 & 0 & 0 & 0 & 0 & 0 & m_s
\end{bmatrix}
$$

The damping matrix can be expressed as

$$
C_i = \begin{bmatrix}
c_i r_{pi}^2 & -c_i r_{pi} r_{gi} & c_i r_{pi}\cos\alpha & -c_i r_{pi}\cos\alpha & c_i r_{pi}\sin\alpha & -c_i r_{pi}\sin\alpha & 0 & 0 \\
-c_i r_{pi} r_{gi} & c_i r_{gi}^2 & -c_i r_{gi}\cos\alpha & c_i r_{gi}\cos\alpha & -c_i r_{gi}\sin\alpha & c_i r_{gi}\sin\alpha & 0 & 0 \\
c_i r_{pi}\cos\alpha & -c_i r_{gi}\cos\alpha & c_{ps} + c_i\cos\alpha & -c_i\cos\alpha & 0 & 0 & -c_{ps} & 0 \\
-c_i r_{pi}\cos\alpha & c_i r_{gi}\cos\alpha & -c_i\cos\alpha & c_{gs} + c_i\cos\alpha & 0 & 0 & 0 & 0 \\
c_i r_{pi}\sin\alpha & -c_i r_{gi}\sin\alpha & 0 & 0 & c_{ps} + c_i\cos\alpha & -c_m\sin\alpha & 0 & -c_{ps} \\
-c_i r_{pi}\sin\alpha & c_i r_{gi}\sin\alpha & 0 & 0 & -c_m\sin\alpha & c_{gs} + c_i\cos\alpha & 0 & 0 \\
0 & 0 & -c_{ps} & 0 & 0 & 0 & c_s & 0 \\
0 & 0 & 0 & 0 & -c_{ps} & 0 & 0 & c_s
\end{bmatrix}
$$

The stiffness matrix can be expressed as

$$
K_i = \begin{bmatrix}
k_i r_{pi}^2 & -k_i r_{pi} r_{gi} & k_i r_{pi}\cos\alpha & -k_i r_{pi}\cos\alpha & k_i r_{pi}\sin\alpha & -k_i r_{pi}\sin\alpha & 0 & 0 \\
-k_i r_{pi} r_{gi} & k_i r_{gi}^2 & -k_i r_{gi}\cos\alpha & k_i r_{gi}\cos\alpha & -k_i r_{gi}\sin\alpha & k_i r_{gi}\sin\alpha & 0 & 0 \\
k_i r_{pi}\cos\alpha & -k_i r_{gi}\cos\alpha & k_s + k_i\cos\alpha & -k_i\cos\alpha & 0 & 0 & -k_s & 0 \\
-k r_p\cos\alpha & k_i r_g\cos\alpha & -k_i\cos\alpha & k_s + k_i\cos\alpha & 0 & 0 & 0 & 0 \\
k r_p\sin\alpha & -k_i r_g\sin\alpha & 0 & 0 & k_s + k_i\cos\alpha & -k_m\sin\alpha & 0 & -k_s \\
-k_i r_{pi}\sin\alpha & k_i r_{gi}\sin\alpha & 0 & 0 & -k_m\sin\alpha & k_s + k_i\cos\alpha & 0 & 0 \\
0 & 0 & -k_s & 0 & 0 & 0 & k_s & 0 \\
0 & 0 & 0 & 0 & -k_s & 0 & 0 & k_s
\end{bmatrix}
$$

The incentive item can be expressed as

$$
P = \begin{bmatrix}
T_{in} - n_{mi}k_{mi}R_{pi}\left(b_i^* - \beta_i\dot{\delta}_{mi}\delta_{mi}\right) \\
T_{out} + n_{mi}k_{mi}R_{gi}\left(b_i^* - \beta\dot{\delta}_{mi}\delta_{mi}\right) \\
n_{mi}k_{mi}\left(b_i^* - \beta\dot{\delta}_{mi}\delta_{mi}\right)\cos\alpha + f_{px} \\
-n_{mi}k_{mi}\left(b_i^* - \beta\dot{\delta}_m\delta_m\right)\cos\alpha + f_{gx} \\
n_{mi}k_{mi}\left(b_i^* - \beta\dot{\delta}_m\delta_m\right)\sin\alpha + f_{py} \\
-n_{mi}k_{mi}\left(b_i^* - \beta\dot{\delta}_m\delta_m\right)\sin\alpha + f_{gy} \\
-f_{px} - f_{gx} \\
-f_{py} - f_{gy}
\end{bmatrix}
$$

Among them $c_i = c_{mi} + \beta_i k_{mi} \left(b_i^* + \delta_{mi} \right)$, $k_i = k_{mi} n_{mi} (1 + \beta \dot{\delta}_m)$, where k_{mi} and c_{mi} are gear meshing stiffness and damping, respectively, and α is the gear meshing angle.

5.6.2 Influence of Friction on the Performance of the Servo System

Friction is a nonlinear link ubiquitous in servo systems. It causes low-speed creeping, increases the residual error of the system, and reduces stability. In the radar antenna-servo system, friction mainly exists at the meshing tooth surface and the bearing, which will be discussed separately below [24, 25].

5.6.2.1 Influence of Gear Meshing Friction
In the process of gear transmission, there is a relative slip between a pair of meshing tooth surfaces, as shown in Figure 5.47.

This relative slip leads to the appearance of meshing friction. Therefore, it can be considered that the meshing friction is mainly manifested as sliding friction, and the magnitude of the frictional force is related to the pressure on the tooth surface. For this reason, the sliding friction torque can also be included in the damping matrix and stiffness matrix in the form of gear meshing force.

5.6.2.2 Influence of Bearing Friction
Similarly, since the friction at the bearing is mainly manifested as rolling friction, studies have shown that the LuGre model can be more realistically reflected in the friction torque at the rotating machinery [26, 27]. Taking the first-stage meshing gear as an example, the tooth backlash, bearing clearance, and friction are introduced into the system dynamics model. The transmission coupling equation is

$$M_i \ddot{X} + C_i \dot{X} + K_i X = P_i \tag{5.41}$$

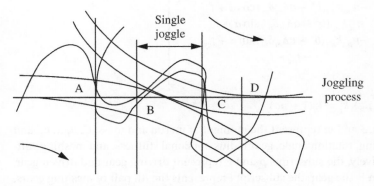

Figure 5.47 Schematic diagram of tooth surface meshing.

Then, the system mass matrix, damping matrix, stiffness matrix, and excitation term are, respectively,

$$M_i = \begin{bmatrix} J_{pi} + \frac{1}{3}J_{spi} & 0 & 0 & 0 & 0 & 0 & 0 & 0 \\ 0 & J_{gi} + \frac{1}{3}J_{sgi} & 0 & 0 & 0 & 0 & 0 & 0 \\ 0 & 0 & m_{pi} & 0 & 0 & 0 & 0 & 0 \\ 0 & 0 & 0 & m_{gi} & 0 & 0 & 0 & 0 \\ 0 & 0 & 0 & 0 & m_{pi} & 0 & 0 & 0 \\ 0 & 0 & 0 & 0 & 0 & m_{pi} & 0 & 0 \\ 0 & 0 & 0 & 0 & 0 & 0 & m_s & 0 \\ 0 & 0 & 0 & 0 & 0 & 0 & 0 & m_s \end{bmatrix}$$

$$C_i = \begin{bmatrix} c_i^f r_{pi}^2 & -c_i^f r_{pi} r_{gi} & c_i r_{pi}\cos\alpha & -c_i r_{pi}\cos\alpha & c_i r_{pi}\sin\alpha & -c_i r_{pi}\sin\alpha & 0 & 0 \\ -c_i^f r_{pi} r_{gi} & c_i^f r_{gi}^2 & -c_i r_{gi}\cos\alpha & c_i r_{gi}\cos\alpha & -c_i r_{gi}\sin\alpha & c_i r_{gi}\sin\alpha & 0 & 0 \\ c_i r_{pi}\cos\alpha & -c_i r_{gi}\cos\alpha & c_{ps}+c_i\cos\alpha & -c_i\cos\alpha & 0 & 0 & -c_s & 0 \\ -c_i r_{pi}\cos\alpha & c_i r_{gi}\cos\alpha & -c_i\cos\alpha & c_{gs}+c_i\cos\alpha & 0 & 0 & 0 & 0 \\ c_i r_{pi}\sin\alpha & -c_i r_{gi}\sin\alpha & 0 & 0 & c_{ps}+c_i\cos\alpha & -c_i\sin\alpha & 0 & -c_s \\ -c_i r_{pi}\sin\alpha & c_i r_{gi}\sin\alpha & 0 & 0 & -c_{mi}\sin\alpha & c_{gs}+c_i\cos\alpha & 0 & 0 \\ 0 & 0 & -c_s & 0 & 0 & 0 & c_s & 0 \\ 0 & 0 & 0 & 0 & -c_s & 0 & 0 & c_s \end{bmatrix}$$

$$K_i = \begin{bmatrix} k_i^f r_{pi}^2 & -k_i^f r_{pi} r_{gi} & k_i r_{pi}\cos\alpha & -k_i r_{pi}\cos\alpha & k_i r_{pi}\sin\alpha & -k_i r_{pi}\sin\alpha & 0 & 0 \\ -k_i^f r_{pi} r_{gi} & k_i^f r_{gi}^2 & -k_i r_{gi}\cos\alpha & k_i r_{gi}\cos\alpha & -k_i r_{gi}\sin\alpha & k_i r_{gi}\sin\alpha & 0 & 0 \\ k_i r_{pi}\cos\alpha & -k_i r_{gi}\cos\alpha & k_{ps}+k_i\cos\alpha & -k_i\cos\alpha & 0 & 0 & -k_s & 0 \\ -k_i r_{pi}\cos\alpha & k_i r_{gi}\cos\alpha & -k_i\cos\alpha & k_{gs}+k_i\cos\alpha & 0 & 0 & 0 & 0 \\ k_i r_{pi}\sin\alpha & -k_i r_{gi}\sin\alpha & 0 & 0 & k_{ps}+k_i\cos\alpha & -k_i\sin\alpha & 0 & -k_s \\ -k_i r_{pi}\sin\alpha & k_i r_{gi}\sin\alpha & 0 & 0 & -k_{mi}\sin\alpha & k_{gs}+k_i\cos\alpha & 0 & 0 \\ 0 & 0 & -k_s & 0 & 0 & 0 & k_s & 0 \\ 0 & 0 & 0 & 0 & -k_s & 0 & 0 & k_s \end{bmatrix}$$

$$P_i = \begin{bmatrix} T_{in} - n_{mi}k_{mi}R_p(b^* - a\dot{\delta}_m\delta_m)(1+\mu_i\varsigma) - F_{pf} \\ T_{out} + n_{mi}k_{mi}R_g(b^* - a\dot{\delta}_m\delta_m)(1+\mu_i\varsigma) - F_{gf} \\ n_{mi}k_{mi}(b^* - a\dot{\delta}_m\delta_m)\cos\alpha + f_{px} \\ -n_{mi}k_{mi}(b^* - a\dot{\delta}_m\delta_m)\cos\alpha + f_{gx} \\ n_{mi}k_{mi}(b^* - a\dot{\delta}_m\delta_m)\sin\alpha + f_{py} \\ -n_{mi}k_{mi}(b^* - a\dot{\delta}_m\delta_m)\sin\alpha + f_{gy} \\ -f_{px} - f_{gx} \\ -f_{py} - f_{gy} \end{bmatrix}$$

$$c_i^f = c_i(1+\mu\varsigma), \quad k_i^f = k_i(1+\mu\varsigma)$$

Among them, J and m represent the moment of inertia and mass, C, θ, R, k_s, and k_m are damping, rotation angle, gear radius, torsional stiffness, and meshing stiffness, respectively, the subscripts p and g represent driving gear and driven gear, respectively, α is the gear, the subscript i represents the ith pair of meshing gears, and T_{in} T_{out} represent the input torque and output torque of the pair of gears.

5.6.3 Construction of Servo System Prototype and Experiment

5.6.3.1 Servo System Prototype

In order to confirm the mechanism and analysis model of the influence of mechanical structure factors on the performance of the servo system, a servo experimental prototype with variable structure parameters is designed and manufactured, which mainly consists of two parts, one is the mechanical structure and the other is digital control system. The mechanical part is a single-axis rotary stage, as shown in Figure 5.48. The platform can change its tooth clearance by adjusting the gear center distance (adjustment range $1'\sim5'$), change the transmission accuracy by replacing the gears (6–7 levels), change the torsional stiffness of the system by changing the drive shaft (resonance frequency can be changed from 9 to 19 Hz discontinuously), and change the load inertia of the system by adjusting the mass block radius of rotation (inertia adjustment range 0.02248–0.22032 kg m^2 and total inertia variation ratio 1–2.314) and the pressure of the friction loading device to obtain different frictional moments (maximum static frictional moment adjustment range 0.2–5 N m).

The change in the structural parameters in the servo test bench is realized by the parameter adjustment mechanism. The parameter adjustment mechanism and

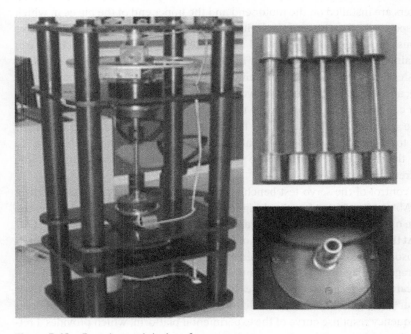

Figure 5.48 Experimental device of servo system.

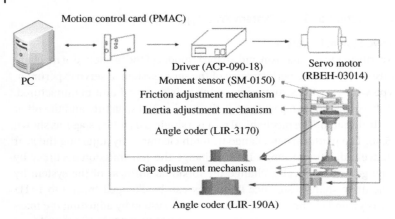

Figure 5.49   Servo test bench and its control system.

the sensor detection part are shown in Figure 5.49. To achieve accurate modeling of the system, the corresponding parameter values before and after structural adjustment should be measurable. Among them, the measurement of the tooth backlash is realized by a high-precision encoder (precision 1.5″) installed on the test bench. Encoders are installed on the motor end and the upper end of the big gear, which are, respectively, fixedly connected with the large and small gears. The motor drive transmission system rotates forward and backward, and the tooth backlash can be obtained through the readings of the two encoders. The torsional stiffness of the system can be calculated by locking the shaft, driving the motor, reading the torque, and the angle difference between the upper and lower encoders. The load inertia can be directly calculated by the geometric parameters of the test bench and the screw of the adjusting mass. The friction torque can be directly measured by the torque sensor (SM-0150) installed on the friction loading mechanism. To ensure the test accuracy of the above parameters, a method of averaging multiple measurements is adopted.

The control of the servo test bench is realized by the two-axis motion control card PMAC-Mini PCI. After the designed control scheme is programmed on the PC, the operation of the test bench can be directly controlled through the PMAC card. At the same time, the operation of the system is measured by the corresponding sensors. Later, it is also transmitted to the PC through the motion control card for storage, display, or closed-loop control. In addition, the system is designed with a dedicated interface, which can be easily connected to a dynamic characteristic analyzer (e.g. Agilent 35670 A) to realize a frequency sweep experiment and obtain the frequency response curve of the experimental platform, which provides a reference for modeling and analysis.

Figure 5.50 Comparison of the speed response simulation and experiment (frequency 5 Hz, voltage 4 V).

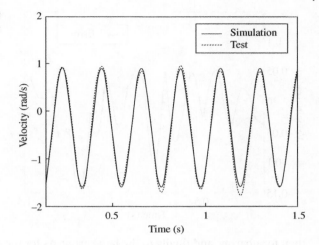

5.6.3.2 Experiment

Based on the abovementioned method, an analysis model of the influence mechanism of the transmission system can be established, and the established model can be verified through an open-loop experiment.

The basic idea of the open-loop experiment is to first obtain the sinusoidal response of the system when nonlinearities are taken into account through simulation and then to obtain the corresponding test results through experiment and compare them to verify the correctness of the established model. Specifically, a sinusoidal excitation signal with a voltage amplitude of 4 V and frequencies of 1, 3, 5, and 7 Hz is applied to the DC servo motor to compare the experimental results with the simulation results. For the sake of space, the analytical and measured values of the system speed response and their error curves are shown in Figures 5.50 and 5.51, respectively, for the 5 Hz case.

Considering the open-loop simulation results and the practical measurement results at different frequencies together, it can be seen that the maximum error between the simulation and the practical measurement does not exceed 0.15 rad/s, and the relative error does not exceed 10%, indicating that the model is correct.

5.7 ISFP of Active Phased Array Antennas with Radiating Arrays

For GBR like active phased array antennas [28] being able to implement wide range of beaming scanning with both machine and electric sweeping, there existed two main random errors, γ_1 and γ_2 in addition to the system error. As far as the random error γ_1 is concerned, it is caused during assembly from the unit, to module,

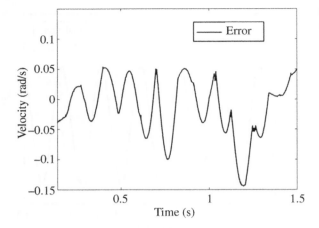

Figure 5.51 Speed response error.

then to subarray, and finally to the large array. As for the random error of γ_2, it is produced during the realization of all-round high-precision movement of the track roller, then to support seat, afterward to achieve a wide range of high-precision pitch axis system, and finally to support the whole array back-frame structure.

Obviously, the array error γ_1 depends on three surfaces, the base support surface, the discrete array element surface, and the spliced conformal surface in between. Naturally, the mathematical characteristics of the errors also depend on a reasonable and concise mathematical description of the random errors on these three surfaces.

5.7.1 Decomposition and Accuracy Transfer of Multilayer Conformal Surfaces

5.7.1.1 Decomposition of Multilayer Conformal Surfaces

To address the root causes of the manufacturing and assembly errors of the active phased array antenna surface and its impact on the antenna performance, a multilayer surface accuracy decomposition model including the base support surface, the spliced conformal surface, and the discrete array element surface is constructed: (i) Base support surface ΔS_1. The task is to realize the support of each subarray, focusing on the influence of cumulative errors in structural error transfer such as truss and beam positioning errors and back frame poses on the support of the array, and then to establish an accuracy characterization model for the base support surface by spinor cluster characterization method. (ii) Spliced conformal surface ΔS_2. The main function is to locate the T/R components, focusing on the impact of errors in wall plate stitching, stitching deformation, heat treatment deformation, and subarray stitching deformation on the accuracy of surface stitching. A hybrid dimensional characterization model is developed to characterize both the integer dimensional errors of the surface and the fractional dimensional errors of the subarrays. (iii) Discrete array element surface ΔS_3. The surface is generally assembled from a large number of discrete T/R components, and the influence of assembly processes such as bolting and welding on the position and pointing deviation of the array elements needs to be considered.

Associative Transfer of Accuracy for Multilayer Conformal Surfaces The geometric accuracy transfer chain is established for a three-layer conformal surface comprising a base support surface, a spliced conformal surface, and a discrete array element surface (Figure 5.52). The first step is to superimpose the deviations of the connection points on the base support surface onto the integer dimensional part of the spliced conformal surface using an associative mapping method and to fit a double cubic B spline to the integer dimensional part of the spliced conformal surface and superimpose it with the fractional dimensional part in order to transfer the geometric deviations of the base support surface to the spliced conformal surface. The second step is to use the projection method to project the discrete array elements in the discrete array surface onto the spliced conformal surface, determine the corresponding points of the discrete array elements in the spliced conformal surface, and obtain the geometric accuracy of the final functional surface by superimposing the coordinates of the corresponding points.

Thus, the random error due to the manufacturing and assembly of the radiation array can be expressed as

$$\begin{aligned}\gamma_1 &= S_0 + \Delta S_1 + \Delta S_2 + \Delta S_3 \\ &= \Delta\varsigma \cdot \vec{n} + (\widetilde{V}^e + \Delta\vec{S}_0)B(u,w) + R_{hd}V_{hd}h_{hd} \\ &\quad + K^{-1}f(S_1, \Delta\vec{S}_2) + T_{m,n} \cdot \Gamma(P_{m,n})\end{aligned} \tag{5.42}$$

where $\Delta S_1 = \left[\Delta S_1^{1,1}, \Delta S_1^{1,2}, \cdots \Delta S_1^{M,N}\right]$, $\Delta S_1^{m,n} = [T_{m,n}, P_{m,n}]^T$, $(m = 1, 2, ..., M)$, $(n = 1, 2, ..., N)$, $\Delta S_2 = p_{hd}(u,w) = p_{id}(u,w) + C_{hd}(u,w) \cdot h_{fd}(u,w)$, $\Delta S_3 = \Delta\varsigma \cdot \vec{n}$, $\widetilde{V}^e$, and $B(u,w)$ are the control vertices and base functions of the bicubic B-sample surface, respectively; $T_{m,n}$ is the local surface rotation of the subarray; $P_{m,n}$ is the transformation vector relative to the position of each rotation; $p_{id}(u,w)$ is the integer dimensional surface component; $C_{hd}(u,w)[w1]$ is the mixed dimensional surface correlation coefficient; $h_{fd}[w1]$ is the fractional dimensional height; ΔS_3 is the projection of the random error $\Delta\varsigma$ in the normal direction $\vec{n}$ of each radial array element due to the accumulation of base support errors; K is the stiffness matrix; and f is the assembly force.

The accuracy of the radial surface can be quantified by surface decomposition and accuracy correlation, which allows the accuracy of the radial surface to be quantified under the effect of different errors such as array support skeleton processing errors, subarray surface splicing errors, and array element assembly attitude errors.

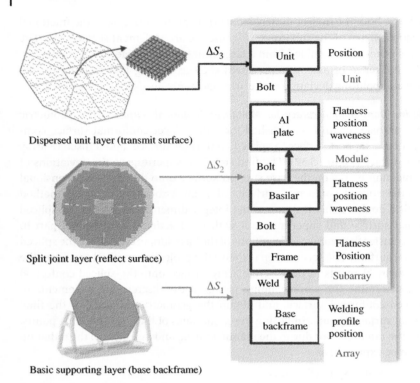

Dispersed unit layer (transmit surface)

Split joint layer (reflect surface)

Basic supporting layer (base backframe)

Figure 5.52 Accuracy-related superposition of multilayered surfaces on functional surfaces.

5.7.2 Accuracy Characterization of Base Support Surfaces

The base support surface itself consists of several subarray positioning surfaces assembled on the support base, and its structure is discrete. It is difficult to completely describe the overall accuracy of the base support surface formed by the multiarray stitching using the existing spin model. To this end, we propose an accuracy characterization model based on a cluster of spins, i.e. to build a spin model for each subarray support surface under a reference system, and to build a relative position transformation vector for each spin according to the relative spatial position of each subarray and merge the spin models under each reference system into the same reference system to describe the discrete geometrical characteristics of the base support surface, thus realizing the accuracy characterization of the positioning surface of the multiarray. The accuracy of the spin cluster is then characterized (Figure 5.53).

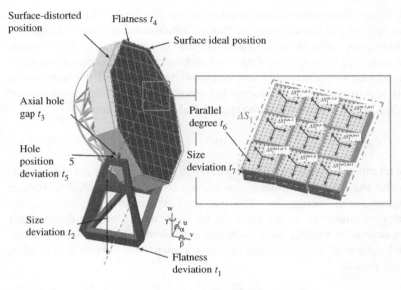

Figure 5.53 Basic support surface characterization.

The accuracy characterization based on the spin-weight cluster model achieves a uniform accuracy characterization of the complete subarray positioning surface, while retaining the independent accuracy information of each subarray. Based on the finite element analysis to obtain the back-frame deflection error T_0 caused by welding deformation, the accuracy transfer model based on Jacobi matrix can be established according to the assembly space relationship of the base, support back frame, and subarray, and the accuracy characterization of the base support surface can be obtained from the base flatness, the height of the support, and the clearance deviation between the support and the axis hole of the array.

5.7.3 Accuracy Characterization of Spliced Conformal Surfaces

The array stitching structure contains both the overall deviations such as the overall array stitching error and detailed errors such as the corrugation of each subarray panel, which are difficult to be described in regular integer dimensions or simply in irregular fractional dimensions. To overcome this difficulty, it is necessary to express the integer dimensional information of the spliced conformal surface as a whole while also accurately expressing the fractional dimensional details of the spliced conformal surface in order to describe the surface accuracy of the spliced conformal surface much clearly and more completely.

(1) The integer dimensional component $p_{id}(u, w)$. The dual cubic B-sample surfaces are used to characterize the integer dimensional surface components, including the subarray splicing accuracy and the overall array deformation. The measured array splicing deviations and the overall array deformation are transformed into a matrix of surface-type value points, and the least square method is used to obtain the control point matrix V of the integer dimensional surface, which is then utilized to create bicubic B-sample surface

$$p_{id}(u, w) = \sum_{i=0}^{m} \sum_{j=0}^{n} V_{ij} N_{i,4}(u) N_{j,4}(w) \tag{5.43}$$

(2) Fractal dimensional component $h_{fd}(u, w)$. A fractal function is used to characterize the fractional dimensional surface components, including surface roughness and corrugation. The fractional dimensional components are expressed uniformly in the same (u, w) space as the integer dimensional components by varying the coordinates and normalizing the A–B function. The height-field function of the fractional dimensional surface is defined in (u, w) space as

$$h_{fd}(u, w) = L \sqrt{\ln \gamma / M} \sum_{m=1}^{M} \sum_{n=0}^{n_{\max}} \gamma^{(D-3)n} \cdot$$

$$\cdot \left(\cos \varphi_{m,n} - \cos \left(\frac{2\pi \gamma^n}{L} \sqrt{(p_{id}(u, w)|_x)^2 + (p_{id}(u, w)|_y)^2} \right. \right.$$

$$\left. \left. \cdot \cos \left(\tan^{-1} \left(\frac{p_{id}(u, w)|_y}{p_{id}(u, w)|_x} \right) - \frac{\pi m}{M} \right) + \varphi_{m,n} \right) \right) \tag{5.44}$$

where $p_{id}(u, w)|_x$ and $p_{id}(u, w)|_y$ are the coordinates of any point in the integer dimension, $h_{fd}(u, w)$ is the height of the point on the fractional dimensional rough surface, D $(2<D<3)$ is the fractional dimension of the rough surface, γ $(\gamma>1)$ is the scale parameter characterizing the spectral density of the rough surface, M is the number of superimposed contour peaks when constructing the surface, $\phi_{m,n}$ is the random phase in $[0, 2\pi]$, $L(L = \max(L_u, L_w))$ is the unidirectional length of the integer dimensional component, n is the number of accumulations, and $n_{max} = \text{int} |\log n_0 / \log \gamma|$, $n_0 = \max(u_s^{-1}, w_s^{-1})$ is the maximum number of samples in one direction of the integer dimensional component.

(3) The mixed dimensional model $p_{hd}(u, w)$. The mixed dimensional influence factor R_{hd} and the mixed dimensional deviation virtual control point V_{hd} can be expressed by the dimensionality of the fractional dimension D, the fractional dimensional roughness G, and the control vertex grid area A_V as

$$\begin{cases} R_{hd} = [G/\sqrt{A_V}]^{(D-2)} \\ V_{hd}(u, w) = \left(\dfrac{P_{wu}(u, w)_x}{|\vec{P}_{wu}(u, w)|}, \dfrac{P_{wu}(u, w)_y}{|\vec{P}_{wu}(u, w)|}, \dfrac{P_{wu}(u, w)_z}{|\vec{P}_{wu}(u, w)|} \right) \end{cases} \tag{5.45}$$

In turn, the mixed dimensional surface correlation coefficient C_{hd} is constructed:

$$C_{hd}(u,w) = R_{hd} \cdot V_{hd}(u,w) \cdot$$
$$\cdot \frac{G^{(D-2)}}{|\vec{P}_{wu}(u,w)| \cdot A_V^{(D/2-1)}} \cdot (P_{wu}(u,w)_x, P_{wu}(u,w)_y, P_{wu}(u,w)_z) \quad (5.46)$$

At this point, the hybrid dimensional model of the spliced conformal surface can be represented as

$$p_{hd}(u,w) = p_{id}(u,w) + C_{hd}(u,w) \cdot h_{fd}(u,w) \quad (5.47)$$

also known as ΔS_2.

5.7.4 Accuracy Characterization of Discrete Array Metasurfaces

In general, the radiation array has large number of discrete dot arrays, usually consisting of multiple subarrays stitched together, and the impact of array element errors in different subarrays on the overall electrical performance is not the same. Traditional sampling methods ignore the variability of the impact of different subarrays on performance and might result in inefficient sampling of the position errors of large-scale discrete array elements and poor accuracy characterization. In order to solve this problem, a block and domain sampling method is proposed for electrical performance amplitude weighting, which can accurately estimate the distribution type and distribution parameters of the geometric errors of discrete array elements in each subarray and then accurately estimate the errors of all discrete array elements in the large radiation array to achieve accurate characterization of discrete array element surfaces, so as to solve the problem that the traditional uniform sampling method can hardly consider the different effects of different subarrays and array elements in different regions on the electrical performance. The method is designed to resolve the difficulty of considering the different effects of different subarrays and different regions of the array on the electrical performance.

References

1 Zhou, J.Z., Huang, J. (2010). Incorporating priori knowledge into linear programming support vector regression. *2010 IEEE International Conference on Intelligent Computing and Integrated Systems (IEEE ICISS2010)*. October 22–24, 2010, Guilin, Guangxi, China.
2 Wang, W., Duan, B.Y., and Ma, B.Y. (2008). A method for testing and adjusting large reflective surface antenna panels and its application. *Acta Electronica Sinica* 36 (6): 1114–1118. (in Chinese).

3 Wei, W., Baoyan, D., and Boyuan, M. (2008). Gravity deformation and best rigging angle for surface adjustment of large reflector antennas. *Chinese Journal of Radio Science* 23 (4): 645–650. (in Chinese).

4 Wei, W., Peng, L., and Liwei, S. (2009). Mechanism of the influence of panel position error on the power pattern of large reflector antennas. *Journal of Xidian University* 36 (4): 708–713. (in Chinese).

5 Wang, W. and Duan, B.Y. (2007). Gravity deformation and gain loss of 12 m reflector antenna. In: *Proceedings of the 2007 IEEE International Conference on Mechatronics and Automation (ICMA 2007)*, August 5–8, 2007, Harbin, China, Vol. 1, 927–931. IEEE.

6 Wang, W., Wang, C.S., and Li, P., et al. (2008) Panel adjustment error of large reflector antennas considering electromechanical coupling. *IEEE/ASME International Conference on Advanced Intelligent Mechatronics (AIM2008)*, August 2-5, 2008, Xi'an, China.

7 Zhou, J.Z., Duan, B.Y., and Huang, J. (2009). Prediction of plane slotted-array antenna electrical performances affected by manufacturing precision. *Journal of University of Electronic Science and Technology of China* 38 (6): 1047–1051. (in Chinese).

8 Zhijian, Y.A.N. (2009). Effect of mechanics structure on the electric performances of radiation element in planar slot antenna. *Telecommunication Engineering* 49 (6): 60–65. (in Chinese).

9 Xiong, C.W. and Wang, Y. (2008). Analytical calculation of the RF equivalent conductivity of the surface of microwave devices. Chinese Institute of Electronics, 1–5 September 2008, Jiujiang, Jiangxi. (in Chinese)

10 Zhou, J.Z., Duan, B.Y., and Huang, J. (2010). Influence and tuning of tunable screws for microwave filters using least squares support vector regression. *International Journal of RF and Microwave Computer Aided Engineering* 20 (4): 422–429.

11 Zhou, J.Z., Huang, J., and Ma, H.B. (2010). Modeling the effect of manufacturing precision on electrical performance of filters using support vector regression. *2010 IEEE International Conference on Mechatronics and Automation (IEEE ICMA2010)*. August 4–7, 2010, Xi'an, China.

12 Jin-zhu, Z., Fu-shun, Z., and Jin, H. (2010). Computer-aided tuning of cavity filters using kernel machine learning. *Acta Electronica Sinica* 38 (6): 1274–1279. (in Chinese).

13 Li-dong, Z. (2009). A study on fabrication of electrically tuned duplex. *Electro-Mechanical Engineering* 25 (5): 46–49. (in Chinese).

14 Shen, Z.F. (2009). ESC duplex filter detection simulation and optimal design. *Defense Manufacturing Technology* 5: 52–55. (in Chinese).

15 Bao-yan, D. (2005). *Analysis, Optimization and Precision Control of Flexible Antenna Structure*. Beijing: Science Press (in Chinese).

16 Li, S.L., Huang, J., and Duan, B.Y. (2010). Integrated design of structure and control for radar antenna servo-mechanism. *Journal of Mechanical Engineering* 46 (1): 140–146. (in Chinese).

17 Huang, J. (2008). Backlash compensation in servo systems based on adaptive backstepping-control. *Control Theory & Application* 25 (6): 1090–1094. (in Chinese).

18 Jinzhu, Z., Baoyan, D., and Jin, H. (2009). Modeling and effects on open-loop frequency for servo systems with backlash. *China Mechanical Engineering* 20 (14): 1721–1725. (in Chinese).

19 Kim, T.C., Rook, T.E., and Singh, R. (2005). Effect of nonlinear impact damping on the frequency response of a torsional system with clearance. *Journal of Sound and Vibration* 281: 995–1021.

20 Hong, B., Duan Baoyan, D., and Jingli. (2008). Simultaneous optimization of control and structural systems of a complex machine. *Chinese Journal of Computational Mechanics* 25 (1): 9–13. (in Chinese).

21 Tong, X.F., Huang, J., and Zhang, D.X. (2009). Multidisciplinary joint simulation technology for servo mechanism analysis. In: *2009 IEEE International Conference on Information and Automation*, vol. 6, 655–658.

22 Huang, J. (2009). HLA based multidisciplinary joint simulation technology for servo mechanism analysis. In: *2009 IEEE International Conference on Mechatronics and Automation*, vol. 8, 4517–4522.

23 Sawalhi, N. and Randall, R.B. (2008). Simulating gear and bearing interactions in the presence of faults. Part I: The combined gear bearing dynamic model and the simulation of localised bearing faults. *Mechanical Systems and Signal Processing* 22 (12): 1924–1951.

24 Jinzhu, Z., Baoyan, D., and Jin, H. (2008). Effect and compensation for servo systems using LuGre friction model. *Control Theory & Application.* 25 (6): 990–994. (in Chinese).

25 Zhou, J.Z. and Huang, J. (2009). Support vector regression modelling and backstepping control of friction in servo system. *Control Theory & Application.* 26 (12): 1405–1409. (in Chinese).

26 Vaishya, M. and Singh, R. (2001). Analysis of periodically varying gear mesh systems with coulomb friction using Floquet theory. *Journal of Sound and Vibration* 243 (3): 525–545.

27 Hensen, R.H.A., Marinus, J.G., and Steinbuch, M. (2002). Frequency domain identification of dynamic friction model parameters. *IEEE Transactions on Control Systems Technology* 10 (2): 191–197.

28 Duan, B. and Tan, J. (2017). Special issue on electromechanical coupling design for electronic equipment. *Chinese Journal of Mechanical Engineering* 30 (3): 495–496.

6

EMC-Based Measure and Test of Typical Electronic Equipment

6.1 Introduction

This chapter investigates the modeling and calculation methods of test factor coupling for typical engineering cases such as airborne planar slotted waveguide array antennas, airborne 3D antenna support with servo system, and the electromodulation duplex filters for scattering telecommunication and addresses the problem of how to determine the main structural factors affecting electrical performance. New test strategies, techniques, and methods are studied to build a semiphysical comprehensive test platform to provide a real basis for verifying the correctness of the theoretical model of electromechanical coupling and the influence mechanism [1].

6.2 EMC-Based Analysis of Measure and Test Factors

The coupling degree among the test factors is the measure of the influence of structural parameters on electrical performance. The purpose of analyzing the coupling degree among test factors is to determine the main mechanical structural parameters affecting the system (electrical) performance in typical cases in order to minimize the testing works.

The technical route of investigating the coupling degree among test factors is shown in Figure 6.1. Firstly, the data envelopment analysis (DEA) method and the subjective scoring method were used, respectively, to calculate the coupling degree among test factors/evaluation weights, and the comprehensive coupling degree among test factors was obtained by combining the subjective and objective methods [2].

Electromechanical Coupling Theory, Methodology and Applications for High-Performance Microwave Equipment, First Edition. Baoyan Duan and Shuxin Zhang.

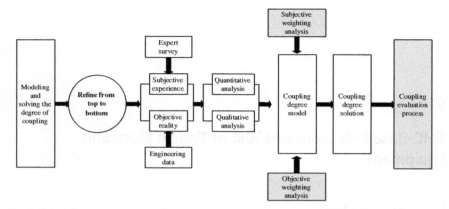

Figure 6.1 Coupling degree analysis among the test factors and the corresponding research roadmap.

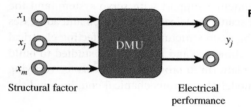

Figure 6.2 DEA decision element.

Structural factor

Electrical performance

6.2.1 Objective Coupling Degree Calculation Method – Data Envelopment Analysis Method

Provided that the input vector of a decision-making element (DMU as in Figure 6.2) in a development activity is $x = (x_1, x_2, \cdots, x_m)^T$ and the output vector is $y = (y_1, y_2, \cdots, y_s)^T$, among the n kinds of $DMU_j (1 \le j \le n)$, the corresponding input and output vectors of DMU_j are

$$x_j = (x_{1j}, x_{2j}, \cdots, x_{mj})^T > 0, j = 1, 2, \cdots, n$$

$$y_j = (y_{1j}, y_{2j}, \cdots, y_{sj})^T > 0, j = 1, 2, \cdots, n$$

and $x_{ij} > 0, \quad y_{rj} > 0, \quad i = 1, 2, \cdots, m, r = 1, 2, \cdots, s$. That is, each decision element has m types of "inputs" and s types of "outputs." x_{ij} is the input of the jth decision element for the ith type of input, and y_{rj} is the output of the jth decision element for the rth type of output.

Let v_i be a measure (weighting) of the input of type i, and u_r be a measure (weighting) of the output of type r. For each decision element DMU_j, the corresponding efficiency evaluation index can be introduced as

$$h_j = \frac{u^T y_j}{v^T x_j} = \frac{\sum\limits_{r=1}^{s} u_r y_{rj}}{\sum\limits_{i=1}^{m} v_i x_{ij}}, \quad j = 1, 2, \cdots, n \tag{6.1}$$

Taking the efficiency index of the j_0th decision element as the objective and the efficiency index of all decision elements as the constraint, the following C^2R model can be constructed as

$$\max \quad h_{j_0} = \frac{\sum\limits_{r=1}^{s} u_r y_{rj_0}}{\sum\limits_{i=1}^{m} v_i x_{ij_0}}$$

$$\text{s.t.} \quad \frac{\sum\limits_{r=1}^{s} u_r y_{rj}}{\sum\limits_{i=1}^{m} v_i x_{ij}} \leq 1 \qquad j = 1, 2, \cdots, n$$

$$v = (v_1, v_2, \cdots, v_m)^T \geq 0$$

$$u = (u_1, u_2, \cdots, u_s)^T \geq 0 \tag{6.2}$$

If Charnes–Cooper method is introduced, it means that

$$t = \frac{1}{v^T x_0}, w = tv, \zeta = tu \tag{6.3}$$

Then, we have the following linear programming problem:

$$\left(P_{C^2R} \right) \begin{cases} \max \ h_{j_0} = \zeta^T y_0 \\ \text{s. t.} \quad w^T x_j - \zeta^T y_j \geq 0, j = 1, 2, \cdots, n \\ w^T x_0 = 1 \\ w \geq 0, \quad \zeta \geq 0 \end{cases} \tag{6.4}$$

The objective coupling degree can be obtained by solving the above programming problem.

6.2.2 Subjective Coupling Degree Calculation Method – Subjective Scoring Method

For some specific evaluation objects, it is often difficult to obtain comprehensive coupling degree information from objective data alone in the process of comprehensive evaluation assignment, which must be combined with subjective experience of experts and professional technicians. Therefore, a subjective scoring method is used to derive the subjective coupling degree after expert consultation by scoring the importance of structural factors relative to electrical performance.

After establishing the evaluation system, the internationally used consulting method of two-by-two comparison of levels 1 to 9 is adopted, and the comparative judgment matrix C of the relative importance of two-by-two factors is obtained (as a positive mutual inverse matrix). Consistency test is conducted on judgment matrix C. If the consistency criterion is satisfied, the judgment matrix C can be used for weight calculation. At this time, the eigenvector w corresponding to the eigenvalue ψ_{max} of the judgment matrix C is the weight of each factor at the lower level to the target at the upper level after normalization.

6.2.3 Combination of Subjective and Objective Coupling Degrees/Weighting

Once the above objective coupling degree vector v and subjective coupling degree vector w are obtained, the following function can be obtained by introducing the weights ϖ_v and ϖ_w:

$$\omega^* = \varpi_v v + \varpi_w w \tag{6.5}$$

where ϖ_v and ϖ_w are the weights for v and w. Determination of ϖ_v and ϖ_w is a combined assignment problem and it can be done by the following two methods.

6.2.3.1 Entropy-Based Metrics

Without loss of generality, let the weighting vector involved in the combination assignment be ω^i and its corresponding combination coefficient be δ_i, then we have

$$\omega^* = \sum_{i=1}^{n} \varpi_i \omega^i \quad i = 1, 2, \cdots, n \tag{6.6}$$

The two-norm of the power vector space is used as the distance metric between ω^* and ω^i,

$$d_i = \|\omega^* - \omega^i\|_2 \tag{6.7}$$

The above optimal weighting problem can be treated as how to make the value of d_i most average in the sense of weight vectors, i.e. how ω^* can combine the effective information of each weight vector as ω^i to the maximum extent. To do this, the information entropy is introduced as a measure of the difference between the sample values of the weightings of the weighting combination system and the relative best combination weightings contained in ω^i, i.e.

$$H = -\frac{1}{\ln n} \sum_{i=1}^{n} p_i \ln p_i \qquad (6.8)$$

where

$$p_i = \frac{\|\omega^* - \omega^i\|_2}{\sum_{i=1}^{n} \|\omega^* - \omega^i\|_2} = \frac{\left\|\sum_{i=1}^{n} \varpi_i \omega^i - \omega^i\right\|_2}{\sum_{i=1}^{n} \left\|\sum_{i=1}^{n} \varpi_i \omega^i - \omega^i\right\|_2} \qquad (6.9)$$

It is easy to know that for p_i there are $0 \leq p_i \leq 1$ and $\sum_{i=1}^{n} p_i = 1$.

6.2.3.2 Measure of the Degree of Coupling Deviation

In the process of combination assignment, both the subjective and objective coupling degrees depend on the data of the research object. When the information contained in the calculated data is not enough, the subjective–objective coupling degree deviates from the "ideal coupling degree." The entropy value as the basis for the measure of the combined coupling degree weighting coefficient ϖ_i does not include the treatment of the above deviation degree. For this reason, U is introduced as the measure of the degree of deviation from the coupling degree as

$$U = \sum_{i=1}^{n} \varpi_i \frac{1}{\sum_{j=1}^{m} \frac{\omega_j^i}{\omega_j^{\max} - \omega_j^{\min}}} \qquad (6.10)$$

where $\omega_j^{\max}$ and $\omega_j^{\min}$ are the maximum and minimum values of the coupling degree vector to be combined, ω_j^i is the value of each coupling degree in the coupling degree vector to be combined, then $\frac{\omega_j^i}{\omega_j^{\max} - \omega_j^{\min}}$ reflects the deviation of this coupling degree vector, and this deviation should be minimized and its inverse value should be maximized in the combination weighting process.

Based on the above analysis, the optimal weighting combination coefficient should maximize the sum of the information entropy and the imbalance measure, i.e.

$$Max\ H = -\frac{1}{\ln n} \sum_{i=1}^{n} p_i \ln p_i + U$$

$$\text{S. T.} \quad \sum_{i=1}^{n} \varpi_i = 1 \qquad (6.11)$$

The optimal combination coefficient ϖ_i is obtained by solving it.

6.3 EMC-Based Measure and Test Technology for Typical Case

6.3.1 Planar Slotted Array Antenna

The planar slotted array antenna (PSAA) is an important equipment for airborne fire-control radar, which contains three layers of cavities, namely, the excitation layer, coupling layer, and radiation layer. The microwave signal is transmitted between two adjacent cavities through the slot coupling form, and the electromagnetic wave is finally radiated to space by the radiation slot of the radiation layer, forming the antenna radiation pattern. The input signal in the excitation layer will be reflected due to the uncompleted matching inside the antenna, forming the input standing wave. High-performance PSAA requires the characteristics of high gain, low sidelobe level, and low voltage standing wave ratio. Figure 3.7 is a physical model of PSAA.

The main structural and electrical parameters of PSAA (Figure 3.7) are as follows:

- Frequency band: X-band
- Structure form:
  900 mm caliber, waveguide wide edge slot
  Three layers of radiation, coupling, and excitation
  Dozens of subarray, more than 1000 radiation slots
- Structural parameters: slot length, slot width, slot offset, size and flatness of each waveguide, etc. for each layer
- Main electrical properties: Gain, sidelobe level, and voltage standing wave ratio.

6.3.1.1 Analysis of the Coupling Degree of PSAA Test Factors

The structure of PSAA is very complex, and there are many structural factors affecting the electrical performance, so it is not practical to test all the structural parameters. This requires the study of the different degrees of influence of the structural factors on the electrical properties, i.e. the degree of test factor coupling, to determine the structural parameters that have much greater impact on the electrical properties and improve the testing efficiency [3, 4].

Calculation System of Coupling Degree According to the coupling degree calculation, combined with the actual PSAA, coupling degree calculation system of PSAA can be established as shown in Figure 6.3.

This makes it possible to clarify the hierarchical relationship between the influence of structural factors on electrical performance and the individual factors contained in each layer of the system, so that the coupling degree can be calculated.

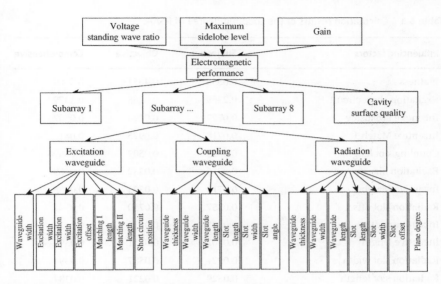

Figure 6.3 Calculation system of coupling degree for PSAA.

Test Factor Coupling Calculation Results and Analysis The test factor coupling of PSAA was obtained using the combined method of subjective coupling and objective coupling (Table 6.1).

After the comprehensive coupling degree calculation, the structural parameters affecting the electrical performance, in the order of importance from largest to smallest, are flatness, excitation layer parameters, coupling layer parameters, and radiation layer parameters, respectively.

6.3.1.2 Key Test Techniques

Existing Test Means for PSAA Table 6.2 shows a summary of the existing instruments used for structural parameters and electrical properties measurements of PSAA [5–13], and a brief description of the new test technique used in this case is given below.

Machine Vision Measurement Technology of PSAA This test technique (Figure 6.4) is employed to perform the contactless testing of the radiation slot shape dimension and waveguide cavity dimension of PSAA and solve the problem of real-time fine dimension measurement of large-diameter PSAA efficiently. Through the measurement software in the machine vision measurement system, the key information is extracted to calculate each measured dimension.

Dynamic Surface Error The existing test means can only test the performance of the PSAA in the not mounted vibration environment, and the vibration level of

Table 6.1 Calculation results of the coupling degree of PSAA.

Influencing factors	Subjective	Objective	Comprehensive
Flatness	0.0555	0.4651	0.2489
Excitation short circuit	0.2369	0.0268	0.1377
Incentive broadside	0.07779	0.0194	0.0502
Incentive Match 1	0.0718	0.0925	0.0816
Coupling slot length	0.0625	0.0507	0.05694
Excitation bias	0.0736	0.0382	0.0569
Incentive Match 2	0.0718	0.0373	0.0555
Radiation slot offset	0.0357	0.0750	0.0542
Incentive width	0.0325	0.0533	0.0423
Incentive length	0.0562	0.0211	0.0397
Radiation slot width	0.0221	0.0587	0.0394
Radiation slot length	0.0529	0.0221	0.0384
Coupling seam inclination	0.0543	0.0201	0.0382
Radiation short circuit length	0.0530	0.0097	0.0326
Radiation waveguide wall thickness	0.0432	0.0097	0.0274

the airborne environment where the planar slotted array antenna works is large, which will not only make the antenna vibrate as a whole but also cause the antenna array to be deformed. In order to achieve real-time online testing, high-speed photogrammetry (Figure 6.5) and accelerometer testing technology (Figure 6.6) can be used, respectively.

Electrical Performance Measurement Technology The electrical performance test is concerned with three indicators, antenna standing wave, sidelobe level, and gain, and it is expected to achieve static and dynamic testing. Figure 6.7 shows a planar near-field test system for radiation pattern measurement, while the standing wave measurement is done by a vector network analyzer.

6.3.2 Three-dimensional Antenna Base Test Technology

Three-dimensional antenna mounts are used in airborne long-range pulsed Doppler fire-control radar, which requires high performance of the servo system, including good dynamic characteristics, high tracking accuracy, large search range, airspace stabilization, and high reliability. The outermost layer of the 3D

Table 6.2 Existing test means for PSAA.

No.	Instrument name	Parameters used for measurement	Main indicators of the instrument	Output form of measurement results
1	Vector network analyzer	Measure antenna standing wave, directional map	R&S ZVK or Agilent 8720ES covering X-band	Data file reports
2	Near field test systems	Measure antenna standing wave, wave flap, gain	Identified as suitable for X-band antenna near-field testing	Data file reports
3	Coordinate measuring machine	The length, width, and position dimensions of waveguide cavity and slit in radiation plate and feed plate; the root mean square value of the array of flat plate antenna	Accuracy $(3.2 + L/250)$ μm	Data file reports
4	Thickness gauge	For flat antenna paint thickness	Paint layer thickness less than 0.1 mm	Data file reports
5	Tensile testing machine	For flat panel antenna strength measurements	$\Sigma b \geq 190$ MPa	Data file reports
6	Thickness gauge	For measuring the thickness of the film layer of antenna components (the same as the paint thickness gauge)	Adhesive layer thickness not more than 0.2 mm	Data file reports
7	Surface resistance tester	Flat antenna conductive oxidation after testing the surface resistance of the coating layer		Data file reports

antenna mount is the cross-roll drive branch, which is connected to the H-shaped frame of the carrier and fixed, followed by the pitch drive branch and the azimuth drive branch, and the azimuth and pitch drive branches are orthogonal. Each degree of freedom constitutes an independent servo drive system. The 3D antenna holder is a complex and high-precision transmission mechanism, including antenna, gimbal, seat, and actuating assembly. The antenna holder shape is shown in Figure 6.8.

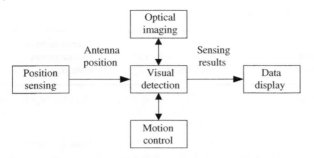

Figure 6.4 Block diagram of machine vision measurement principle.

Figure 6.5 High-speed photogrammetry.

The main structural parameters and servo performance parameters of the 3D antenna mount are as follows:

- Structure form: three axes of azimuth, pitch, and horizontal roll
- Structural parameters: friction torque, backlash, clearance, modalities, orthogonality, etc.
- Main servo performance: pointing accuracy, overshoot, rise time, adjustment time, and Servo bandwidth.

6.3.2.1 Three-dimensional Antenna Base Test Factor Coupling Degree Analysis

The structure of 3D antenna mounts is very complex, and there are many structural factors affecting the electrical performance, and it is extremely difficult to

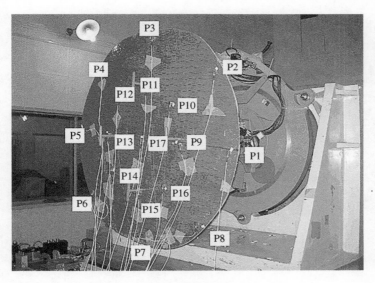

Figure 6.6 Acceleration sensor measurement.

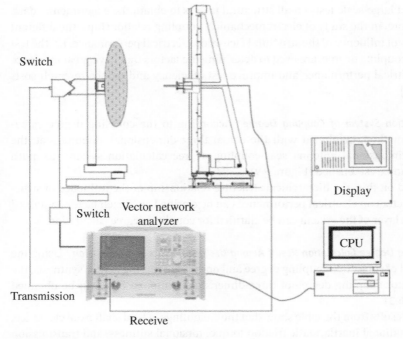

Figure 6.7 Planar near-field test system.

Figure 6.8 Schematic diagram of three-dimensional antenna base structure.

conduct large-scale tests on all structural factors to obtain the experimental data. Therefore, in the study of electromechanical coupling relationships, the different degrees of influence of the structural factors on electrical performance, i.e. the test factor coupling degree, are used to determine the factors that have great influence on electrical performance and improve test efficiency and reduce research costs [14, 15].

Calculation System of Coupling Degree According to the coupling degree calculation method, combined with the actual three-dimensional antenna seat, the three-dimensional antenna seat coupling degree calculation system (azimuth axis) can be established (Figure 6.9).

Based on this, the hierarchical structure relationship of the influence of structural factors on electrical performance can be clarified, and each factor contained in each layer of the system can be clarified for coupling degree calculation.

Coupling Degree Calculation Result Among the Test Factors and Discussion Using the method of subjective coupling degree and objective coupling degree synthesis, the test factor coupling degree of three-dimensional antenna base can be obtained (Table 6.3).

The results from the table show that the weightings of load shaft axial clearance, load rotational inertia, static friction torque, torsional stiffness, and transmission return difference are relatively large in coupling degree, while the weightings of

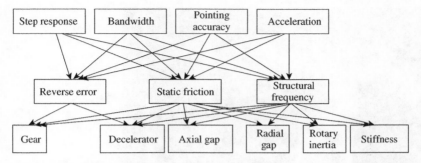

Figure 6.9 Three-dimensional antenna base orientation drive chain coupling degree calculation system.

Table 6.3 Calculation results of azimuthal coupling degree.

Parameters	F_r	F_w	f_{pb}	f_f	E_w	R_d	S_{ft}	L_{sac}
Subjective	0.0643	0.0471	0.0471	0.0471	0.0471	0.1246	0.1246	0.0617
Objective	0.0129	0.0299	0.0168	0.0339	0.0237	0.1378	0.1678	0.2882
Comprehensive	0.05169	0.0429	0.03967	0.04388	0.04135	0.1279	0.13525	0.117456
Parameters	B_{12}	B_{13}	B_{14}	B_{15}	T_s	E_s	R_i	
Subjective	0.0172	0.0172	0.0172	0.0172	0.1439	0.0314	0.1919	
Objective	0.0179	0.0044	0.0043	0.0034	0.0511	0.0647	0.1429	
Comprehensive	0.0174	0.0141	0.0141	0.0138	0.1211	0.0396	0.1799	

drive chain large gear and drive shaft inner and outer diameter dimensions are relatively small.

6.3.2.2 Key Test Techniques

Three-axis Orthogonality Test The three-axis orthogonality test has two test methods: laser and downscaled asymptotic. In order to solve the three-axis orthogonality test problem brought by the three-dimensional antenna base with three degrees of freedom linkage, the three-dimensional descending asymptotic measurement method from the inner ring to the outer ring is adopted (Figure 6.10), which can reduce the test difficulty while ensuring the accuracy [16].

Bearing Axial Clearance Test Due to the limitation of testing space, the axial clearance test adopts the plug gauge comparison method. Between the two split outer rings of the bearing, the plug gauge is gradually added until the bearing

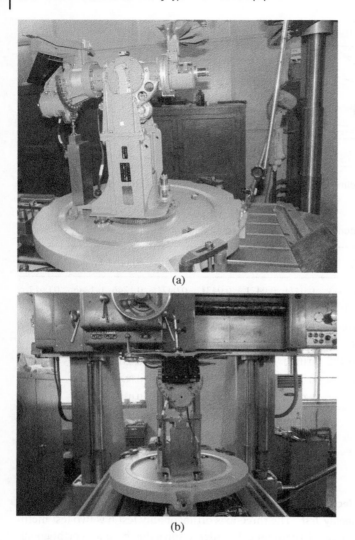

(a)

(b)

Figure 6.10 Three-dimensional antenna base orthogonality test site.

rotates flexibly, and then the table is listed to measure the axial clearance of the bearing (Figure 6.11).

Gear Returning Test According to the characteristics of the antenna base structure, the transmission return test is performed by taking a table on the large gear. When the gearbox is locked, the gear is rotated. According to the measured data, the

Figure 6.11 Three-dimensional antenna mount split bearing axial clearance test.

return value is calculated. The method is not only simple and accurate but also easy to operate.

Gear Form Tolerance Test Special testing equipment for gear form tolerance (gear measuring center, gear runout checker, and common normal micrometer) is used to test the load gear form tolerance.

Low-temperature Drive Chain Friction Torque Test In the low-temperature state, the method of external adjustable power supply to the low-temperature test chamber on the drive chain motor winding energized (with the drive chain angle sensor angle test to observe the load movement) is used to calculate the winding current and convert it to the drive chain friction torque. The low-temperature friction torque test is shown in Figure 6.12.

Servo Performance Test Servo performance test includes step response quality index (response rise time t_r, adjustment time t_s, overshoot δ, and oscillation number n) test, servo bandwidth test, antenna pointing accuracy test, and angular acceleration test. The test system composition is shown in Figure 6.13.

6.3.3 Electrically Tuned Duplex Filter Test Technique

Electrically tuned duplex filters, as the important electromagnetic wave-processing functional parts, are mainly used in scattering telecommunication equipment. Figure 6.14 shows the 3D modeling of the four-cavity structure of the ESC duplex

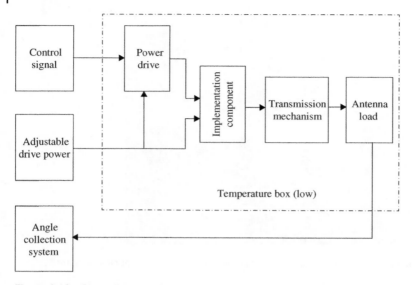

Figure 6.12 Block diagram of low-temperature drive chain friction torque test.

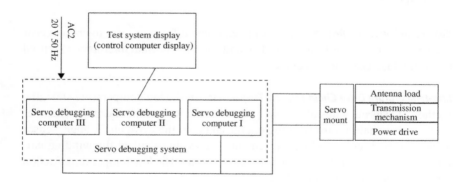

Figure 6.13 Block diagram of 3D antenna mount servo performance test.

filter, with an inner conductor inside the cavity for tuning the operating frequency and coupling holes between the cavities to realize the coupling of electromagnetic waves between the cavities, and the input and output of electromagnetic waves of the ESC duplex filter through the coupling ring.

The main structural parameters and electrical performance parameters of the electrically tuned duplex filter are as follows:

- Frequency band: 610–960 MHz
- Structure: Coaxial cavity-type sequential coupling

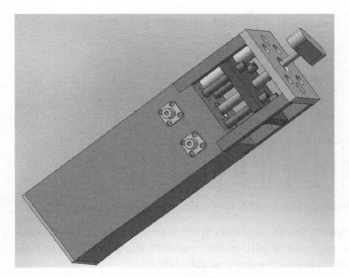

Figure 6.14 Schematic diagram of ESC duplex filter structure.

- Structure parameters: coupling ring, coupling hole shape, size; inner conductor size; shape cavity size, etc.
- Main electrical properties: center frequency, bandwidth, insertion loss, return loss, and stopband rejection.

6.3.3.1 Electrically Tuned Duplex Filter Test Factor Coupling Degree Analysis
The structural factors affecting the electrical performance index of the filter are the coupling hole size and shape, position accuracy, coupling ring size and shape, position accuracy, inner conductor length and accuracy, and cavity equivalent outer diameter to inner conductor diameter ratio. The coupling degree between the key structural factors and the electrical performance index can be obtained by calculating the coupling degree of electromechanical coupling test factors and then determining the key structural factors affecting the electrical performance [17].

Coupling Degree Calculation System According to the coupling degree calculation system establishment method, combining with the actual electrically tuned duplex filter, the coupling degree calculation system can be established as shown in Figure 6.15.

Since the operating frequency of this filter is low and the surface accuracy obtained by the existing processing method can basically meet the electrical performance index requirements, the influence of the two factors of the cavity surface roughness and the surface coating thickness can be ignored. Through the

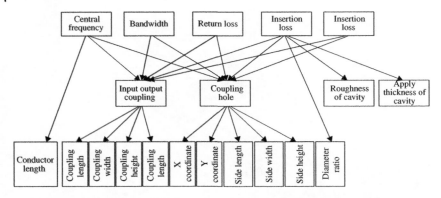

Figure 6.15 The electrically tuned duplex filter coupling calculation system.

above calculation system, the hierarchical structure relationship of the influence of structural factors on electrical performance and each factor contained in each layer can be clarified for the coupling degree calculation.

Test Factor Coupling Calculation Results and Discussion Using a combination of subjective and objective coupling, the test factor coupling of the ESC duplex filter can be obtained (Table 6.4).

After comprehensive coupling analysis and calculation, in the factors affecting the electrical performance, the coordinates of coupling hole position Y_0 and X_0, the ratio of conductor equivalent outer diameter to inner conductor diameter D/d, and the processing accuracy of inner conductor length are secondary factors and may not be tested.

6.3.3.2 Key Test Techniques [18, 19]

Coupling Hole Size and Shape, Position Accuracy Test The coupling hole is located in the cavity partition as a trapezoidal structure. The required detection data are the coordinates x_0 and y_0 of the coupling hole position, and the long side, short side, and height of the coupling hole are L_{ka}, L_{kb}, and h. Before the cavity is welded and formed, all the above data are the detection of the length quantity, which can be achieved using the mature detection method. The main instrument is the universal tool microscope with a measurement accuracy of 2 μm. After the cavity is welded and formed, the coupling hole is difficult to measure. For this reason, a CMM is used for measurement with an accuracy better than 2 μm.

Coupling Ring Size and Shape, Position Accuracy Test The coupling ring is mounted in the cavity, and the dimensions of the coupling ring after molding can be measured directly. The dimensions that have a correlation with the electrical performance index are the long side of the coupling ring L_{ha}, the short side of the

Table 6.4 Calculation results of coupling degree of electrically tuned duplex filter.

Parameters	Coupling ring, short side	Coupling hole, short side	Coupling entrance/exit to short road distance	Coupling ring, height	Coupling ring, long side	Coupling hole height
Subjective	0.2656	0.1689	0.1113	0.1033	0.0846	0.08164
Objective	0.0941	0.0761	0.00239	0.6841	0.0474	0.001801
Comprehensive	0.2097	0.1386	0.07581	0.2926	0.07256	0.05561
Parameters		Coupling hole position coordinates Y_0	Coupling hole position coordinates X_0	Conductor equivalent outer diameter to inner conductor diameter ratio D/d	Length of inner conductor processing accuracy	
Subjective	0.08164	0.0539	0.02878	0.00232	0.01775	
Objective	0.02608	0.00802	0.04723	0.01106	0.00159	
Comprehensive	0.06352	0.03896	0.0348	0.00517	0.01247	

coupling hole L_{hb}, the height of the coupling ring L_{hh}, and the distance H from the coupling ring to the short road surface. The main inspection instrument is a universal tool microscope.

Inner Conductor Length Test The inner conductor length dimension is the working length of the inner conductor in the cavity, i.e. the resonant length, which is a variable in the full-band tuning process (600 working points) and mainly affects the center frequency. To be able to reflect the actual working length in relation to the center frequency, the working state of the filter needs to be detected, and here the height meter is used, i.e. the length dimension of the inner conductor in the cavity is measured directly.

Cavity Equivalent Outer Diameter to Inner Conductor Diameter Ratio Test The ratio of cavity equivalent outer diameter to inner conductor diameter, i.e. the ratio of each inner cavity profile perimeter dimension to the cavity conductor diameter, is the measurement of the length quantity. In the filter assembly process, the cavities and inner conductors with close ratio are assembled on a filter as much as possible. Measurements are made using internal and external micrometers.

Deep Cavity Dimension Measurement The measurement of the dimensions in the deep cavity is difficult and critical for testing. Since the filter assembly is a thin-walled deep cavity structure, in order to prevent scratches and clamping deformation, Swift flexible fixture is used for clamping and partially fixed using a special method of pasting. The coordinate system is established with the bottom of the cavity as the reference. A carbon stylus with the diameter of 5 and 200 mm extension of the SP25-3 module is used for measurement during scanning. The stylus joint with 0.7 mm diameter spherical stylus is selected for dimensional inspection, which can avoid collision when scanning long distances in a small space. In the 3D software, the cross-sectional contour lines are formed based on the point cloud data to realize the dimensional measurement in the deep cavity. Figures 6.16 and 6.17 show the physical photos of some of the main test equipment.

Electrical Performance Test The electrical performance test of the ESC duplex filter is performed with vector network analyzer. Since the specific test methods are relatively mature, they are not repeated.

6.4 EMC-Based Measure and Test System for Typical Case

On the basis of solving the key test techniques of typical cases, a comprehensive test platform can be built to realize the integrated test of structural and electrical

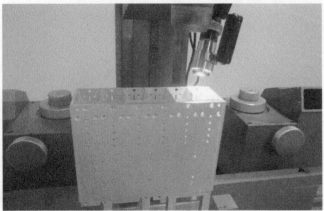

Figure 6.16 Heavy-duty tool microscope measuring straight.

performance in typical cases and provide the basis for comprehensive evaluation. The comprehensive test system has the following advantages:

(1) Testing efficiency can be, seeing from macro point of view, improved by the unified platform.
(2) It is easy to realize the unification of the different test data formats of different test equipment. At the same time, the test data is unified into the database, which can not only realize the data sharing but also facilitate data management.
(3) It facilitates the implementation and evaluation system interface to form a unified comprehensive test and evaluation platform system.

Figure 6.17 CMM measurement of internal cavity dimensions.

6.4.1 Planar Slotted Array Antenna-integrated Test Platform

The comprehensive test platform of PSAA is shown in Figure 6.18.

The platform mainly integrates the dynamic surface errors, internal defects, shape dimension, deformation test, and electrical performance test of PSAA. The test data measured by this comprehensive test platform are uniformly saved in the corresponding database to form the complete original test data corresponding to the product, which is shared by the electromechanical coupling comprehensive evaluation system [20–23].

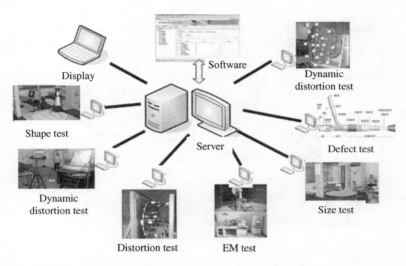

Figure 6.18 Planar slotted array antenna-integrated test platform.

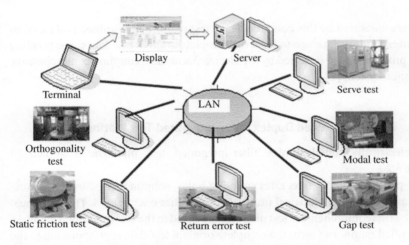

Figure 6.19 Three-dimensional antenna base-integrated test platform.

6.4.2 Three-dimensional Antenna Base-integrated Test Platform

The three-dimensional antenna base comprehensive test platform is shown in Figure 6.19.

The platform mainly integrates antenna seat modal, axial clearance, transmission return, static friction, three-axis orthogonality, and servo performance tests.

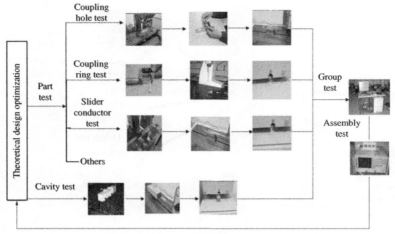

Data feedback from actual EM test

Figure 6.20 Electrically tuned duplex filter-integrated test platform.

The data measured by this comprehensive test platform are unified and saved in the corresponding database to form the complete original test data corresponding to the product, which is shared by the electromechanical coupling comprehensive evaluation system [24].

6.4.3 Electrically Tuned Duplex Filter-integrated Test Platform

The electrically tuned duplex filter-integrated test platform is shown in Figure 6.20.

The platform mainly integrates coupling holes, coupling rings, inner conductors, cavity structure, electrical properties, and other energy tests. The data measured by this comprehensive test platform are saved to the corresponding database in a unified manner to form the complete original test data corresponding to the product, which is shared by the comprehensive electromechanical coupling evaluation system [25, 26].

References

1 Xian-rong, S.H.U., Bing-fa, H.E., and Xian-song, G.U.O. (2008). On mechatronics and mechatronic coupling in SAPR. *Electro-Mechanical Engineering* 23 (6): 57–60. (in Chinese).

2 Hongbo, M. and Guangda, C. (2009). The method for computing coupling degree of testing factors in electromechanical problems. *Proceedings of the Annual Conference of the Chinese Institute of Electronics and Mechanical Engineering Branch*, Taiyuan, China, November 2009. (in Chinese).

3 Yu Wei, G. and Wei-jun, G.X.-s. (2008). Pattern analysis for deformed subarray of planar slotted antenna array. *Modern Radar* 30 (12): 70–73. (in Chinese).

4 Qingfeng, S. and Xiansong, G. (2009). Generalized matching equations for slotted-waveguide array antenna impedance match. *Modern Radar* 31 (12): 70–72. (in Chinese).

5 Gui-Ping, H. (2005). Study on the key technologies of digital close range industrial photogrammetry and applications. Tianjin: Tianjin University. (in Chinese).

6 Gui, Z.J.-I. and Sheng-hua, Y.E. (2005). The key technique study of precise coordinate measurement based on digital close range photogrammetry. *Acta Metrologica Sinica* 26 (3): 207–211. (in Chinese).

7 Chin, T.T. and Harlow, C.A. (1982). Automated visual inspection a survey. *IEEE Transactions on Pattern Analysis and Machine Intelligence* 4 (6): 557–573.

8 Fraser, C.S. (1997). Digital camera self-calibration. *Photogram and Remote Sensing* 52 (4): 149–159.

9 Hong, S. (2006). *Modern Digital Image Processing*. Beijing: Publishing House of Electronics Industry (in Chinese).

10 Soini, A. (2001). Machine vision technology take-up in industrial applications. *Image and Signal Processing and Analysis* 332–338.

11 Ross, W.A. (2003). The impact of next generation test technology on aviation maintenance. *IEEE Autotestcon 2003*. Sep. 22–25, CA, USA.

12 Sheng-hong, F.A.N. (2006). Subpixel accuracy artificial target location using canny operator. *Journal of Geomatics Science and Technology* 2: 76–78. (in Chinese).

13 Hua-chao, Y.A.N.G. and Ka-zhong, D.E.N.G. (2006). Decomposing cameras exterior orientation elements using 2D DLT and collinearity equation. *Journal of Geomatics Science and Technology* 6: 232–234. (in Chinese).

14 Jun-Dong, S., Qian, Y., and Li-hao, P. (2006). Study of electromechanical coupling problem of three-dimensional antenna base. *The First Conference on Fundamental Problems of Electromechanical Coupling of Electronic Equipment*, November 10-11, 2006, Chengdu, Sichuan.

15 Yong Jun, D. (2008). Research on eletromechanical integrative simulation of radar antenna pedestal, *Second Conference on Fundamental Problems of Electromechanical Coupling of Electronic Equipment*, May 24–25, 2008, Nanjing, Jiangsu Province of China.

16 Zhengda, L.I., Guangda, C.H.E.N., and Hongbo, M.A. (2008). Structure factor to radar servo system performance influence and its test research. *Modern Electronics Technique* 31 (3): 33–36. (in Chinese).

17 Zhen-Fang, S. (2006). Electromechanical coupling analysis of electrically tuned duplex filter test factors. *The First Conference on Fundamental Problems of Electromechanical Coupling of Electronic Equipment*, November 10-11, 2006, Chengdu, Sichuan of China.

18 Ma, H.B., Yang, D.W., and Zhou, J.Z. (2010). Improved coupling matrix extracting method for Chebyshev coaxial-cavity filter. *Progress in Electromagnetic Research Symposium*, March 22-26, 2010, Xi'an, China. (in Chinese).

19 Ya-Qing, S. (2006). Research on electromechanical coupling test methods for Electrically Tuned filters *The First Conference on Fundamental Problems of Electromechanical Coupling of Electronic Equipment*, November 10–11, 2006, Chengdu, Sichuan of China.

20 Chao-guang, J.I.A., Zhan-jie, T.A.O., and Shuang, Z.H.A.N.G. (2009). On dual cameras synchronization for high speed photogrammetry system. *Journal of Geomatics Science and Technology* 25 (6): 76–78. (in Chinese).

21 Ross, B. (2004). DoD Automatic test systems strategy refresh. *IEEE Int. Auto Test Conference 2004*. Sep. 20–23, CA, USA.

22 Anderson, J.L. Jr., (2003). High performance missile testing. *IEEE Int. Automatic Test Conference 2003*. Sep. 22–25, CA, USA.

23 Hong-lei, Q.I.N. (2003). Design on automatic test system of airborne radar. *Journal of Test and Measurement Technology* 17 (03): 54–58. (in Chinese).

24 Gu, J.F., Ping, L.H., and Liu, J.C. (2008). Coupled vibration analysis for test sample resonating with vibration shaker. *Third Asia International Symposium on Mechatronics (AISM2008)*, August 27–31, Sapporo, Japan.

25 Li-dong, Z.H.E.N. (2009). A study on fabrication of electrically tuned duplex. *Electro-Mechanical Engineering* 25 (5): 46–49. (In Chinese).

26 Zhen-Fang, S. (2009). Electrically tuned duplex filter detection simulation and optimization design. *Defense Manufacturing Technology* 10 (5): 52–59. (in Chinese).

7

Evaluation on EMC of Typical Electronic Equipment

7.1 Introduction

The evaluation of electromechanical coupling (EMC) of electronic equipment includes two parts: the validation of the correctness and the evaluation of the effectiveness of the coupling theory and influence mechanism (IM) in which the validation of the correctness is obtained by comparing the theoretical calculated electrical performance with the measured electrical performance for a typical case [1]. The effectiveness evaluation is for typical cases to see the application of EMC theory and influence mechanism in the equipment. With the application, the effects on electrical performance improvement and cost reduction are evaluated, including the evaluation of the manufacturability and electrical performance evaluation. The evaluation of EMC of electronic equipment will provide a support for the promotion and application of EMC theory.

7.2 On Correctness of EMC Theory and IM

The basic idea of correctness verification is classified into two points, on the one hand, the procedure is to conduct actual tests of structural parameters and electrical performance indexes for typical engineering cases to form a sample database of structural parameters and electrical performance; on the other hand, the procedure is to substitute the measured structural parameters into the coupling theory and influence mechanism for theoretical calculations to obtain the corresponding electrical performance and form the calculation results of coupling theory and influence mechanism. Here, the hypothesis testing method is applied to compare the measured data with the calculated data of coupling theory and influence mechanism to verify the approximation of the calculated and measured values of electrical performances of coupling theory and influence mechanism.

Electromechanical Coupling Theory, Methodology and Applications for High-Performance Microwave Equipment, First Edition. Baoyan Duan and Shuxin Zhang.
© 2023 The Institute of Electrical and Electronics Engineers, Inc. Published 2023 by John Wiley & Sons, Inc.

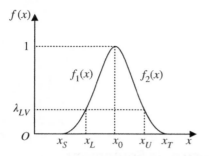

Figure 7.1 Affiliation function curve.

The classical hypothesis testing theory requires that the test samples must obey the normal distribution and often requires sufficient samples, which is objectively limited [2]; for this reason, the original test method can be improved by applying knowledge of fuzzy mathematics and the gray distance measurement.

7.2.1 Fuzzy–Gray Integrated Test Method

7.2.1.1 Gray Estimation as the Estimation of Overall True Value
To accommodate the small number samples, the gray estimation can be used to represent the estimation of the overall true value. The detailed procedure is as follows.

The Determination of Minimum Sample Number and Fuzzy Affiliation Curve Set the total sample data $X = (x_1, x_2, \cdots x_i, \cdots, x_n)$, x_i is the ith measured value, and n is the total number of samples. The affiliation function about x is assumed as

$$f(x) = \begin{cases} f_1(x) & x \leq x_0 \\ f_2(x) & x \geq x_0 \end{cases} \tag{7.1}$$

In order to form a fuzzy affiliation curve as shown in Figure 7.1 (the curve is centered on the middle and has the characteristics of high center and low values on both sides), at least three data points are required. However, considering that in practical engineering, only three data points cannot form the curve shown in the figure, and in the extreme cases, all three data points are on one side of the curve, only half of the curve can be formed. Therefore, the minimum number of data points is determined as 5 to ensure that a complete fuzzy affiliation curve can be formed. In the following study, the minimum number of samples required in the subsequent determination process is determined as 5.

The true value X_0 is estimated by the principle of maximum affiliation, and it is estimated as

$$X_0 \approx x \big|_{f(x)=1} = X_v \tag{7.2}$$

The estimation of the total distribution interval is $x \in [x_L, x_U]$, where x_L and x_U can be expressed by horizontal intercept set λ_{LV} as

$$\begin{cases} x_L = x \,|f_1(x) = \lambda_{LV} \\ x_U = x \,|f_2(x) = \lambda_{LV} \end{cases} \tag{7.3}$$

where λ_{LV} is the optimal level, and $\lambda_{LV} \in [0, 1]$. In the fuzzy set sense, $\lambda_{LV} = 0.5$. Generally, when n is a finite value, $\lambda = 0.4\sim0.5$. If n is smaller, $\lambda_{LV} = 0.4$.

Determination of Gray Estimates For the given sample data $X = (x_1, x_2, \cdots x_i, \cdots, x_n)$, if the coarse errors of the sample data have been removed, each sample data at this time reflects the properties of the true value of the data. Its gray estimation is that

$$\hat{x} = \sum_{i=1}^{n} w_i \cdot x_i \tag{7.4}$$

where w_i is the weight of the sample point x_i in the gray estimation $\hat{x}$, and it satisfies $w_i \geq 0$ and $\sum_{i=1}^{n} w_i = 1$. When each sample point contributes equally to the gray estimation, we have $w_i = 1/n$. In this situation, $\hat{x}$ obtained at this point is consistent with the maximum-likelihood criterion.

If we take each sample point as the comparison data and calculate its gray distance measure with the sample points in the whole sample space separately [2, 3], we can get n gray distance measures. Summing them and taking the average, it is expressed as

$$J_i = \frac{1}{n} \left(\sum_{j=1}^{n} dr(x_i, x_j) \right) \tag{7.5}$$

where

$$dr(x_i, x_j) = \frac{\xi \|d(X, x_i)\|_\infty}{|x_j - x_i| + \xi \|d(X, x_i)\|_\infty} \tag{7.6}$$

$$\|d(X, x_i)\|_\infty = \max_k \{ |x_k - x_i| \quad |k = 1, 2, \cdots, n \} \tag{7.7}$$

ξ is the resolution factor, which is generally taken as 0.5 in engineering, and $J_i (i = 1, 2, \cdots n)$ denotes the gray distance measure of the sample point x_i to the whole sample space.

7.2.1.2 Conversion of Affiliation Order to Affiliation Function

If the gray estimations are considered to be subjected to the overall distribution of the sample data, they can be added to the sample data X, and the new series can be obtained by arranging them in an ascending order:

$$X' = \{x_1, x_2, \cdots, x_v, \cdots, x_{n+1}\} \tag{7.8}$$

Let $x_v = \hat{x}$; then, the order of affiliation is converted into an affiliation function as

$$m_j = \frac{\ln(n + 3 - r(x_j))}{\ln(n + 3)} \quad j = 1, \cdots, (n + 1) \tag{7.9}$$

7.2.1.3 Determination of x_L and x_U [4]

If $f_{1j}(x_j) = m_j \quad j = 1, 2, \cdots, v \tag{7.10}$

$$f_{2j}(x_j) = m_j \quad j = v, v + 1, \cdots, n - 1 \tag{7.11}$$

one can get the following discrete affiliation functions of $f_{1j}(x)$ and $f_{2j}(x)$ by satisfying the interval $[0, 1]$:

$$f_{1j} = f_{1j}(x) = 1 + \sum_{l}^{L} a_l (X_0 - x)^l \tag{7.12}$$

and

$$f_{2j} = f_{2j}(x) = 1 + \sum_{l}^{L} b_l (x - X_0)^l \tag{7.13}$$

In general, taking $L = 3$ to approximate the discrete values $f_{1j}(x_j)$ and $f_{2j}(x_j)$, and let

$$r_{1j} = f_1(x_j) - f_{1j}(x_j) \quad j = 1, 2, \cdots, v \tag{7.14}$$

$$r_{2j} = f_2(x_j) - f_{2j}(x_j) \quad j = v, v + 1, \cdots, n - 1 \tag{7.15}$$

choose $a_l = a_l^*$ and $b_l = b_l^*$ to satisfy the following equations, respectively:

$$\min \|r_1\|_2, \quad \min \|r_2\|_2 \tag{7.16}$$

the constraints are

$$f_1' = \frac{df_1}{dx} \geq 0 \quad f_2' = \frac{df_2}{dx} \leq 0 \tag{7.17}$$

then the predetermined coefficients a_l and b_l, thus, $f_1(x)$ and $f_2(x)$ can be easily obtained. Afterward, x_L and x_U can be known by

$$\min |f_1(x) - \lambda_{LV}|_{x=x_L} \quad \min |f_2(x) - \lambda_{LV}|_{x=x_U} \tag{7.18}$$

Finally, the fuzzy affiliation curve corresponding to the sample data sequence X can be obtained.

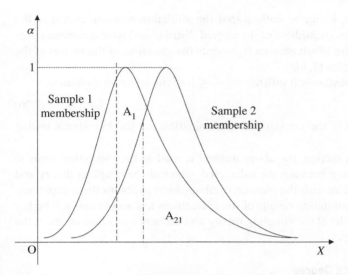

Figure 7.2 Distribution of affiliation functions for the two comparison samples.

7.2.1.4 Two Overall Mean Hypothesis Tests

Suppose that two sequences of data, having been arranged in the ascending order, are $X_j = \{x_{j1}, x_{j2}, \cdots, x_{jn_j}\}$ ($j = 1, 2$). Following the above Sections 7.2.1.1–7.2.1.3, two fuzzy affiliation curves are obtained as shown in Figure 7.2, and the hypothesis testing problem is assumed as $H_0 : \overline{x_1} = \overline{x_2}$.

In Figure 7.2, A_1 is the area of the right-hand part of the sample curve 1, and A_{21} is the area of the part of curve 1 in the right part of the intersection of sample curve 1 and sample curve 2.

The rejection domain solution procedure for the hypothesis test can be determined as follows.

Building the affiliation function of the sample data X_i, the affiliation function with respect to x_{jv} ($i \neq j$, $i = 1, 2$) becomes

$$m(x) = \begin{cases} 0 & x < x_{jL} \\ \dfrac{x - x_{jL}}{x_{jv} - x_{jL}} & x \in [x_{jL}, x_{jv}] \\ \dfrac{x_{jU} - x}{x_{jU} - x_{jv}} & x \in [x_{jv}, x_{jU}] \\ 0 & x > x_{jU} \end{cases} \tag{7.19}$$

It can be seen that the affiliation degree $m(x_{iv})$ corresponding to x_{jv}, $i \neq j$ can be obtained. The above affiliation function becomes an inclusion point analysis for it includes $\hat{x}$, i.e. $\hat{x}$ can be taken as a data of the system.

From Figure 7.2, it can be noticed that the affiliation function curves of the two control samples (regardless of the obeyed distribution) have a common area (Figure 7.2), the size of whose area represents the closeness of the means of the two groups of samples [5, 6].

Let $\delta_A = {}^{A_{21}}/_{A_1}$, obviously it satisfies $0 \leq \delta_A \leq 1$, so the rejection domain is

$$\delta_A \leq \delta_R \tag{7.20}$$

where $0 \leq \delta_R \leq 0.5$ is the constant of determination for the hypothesis testing problem.

In the following section, the above method is used as the theoretical basis to define the agreement between the calculated values of the coupling theory and influence mechanism with the measured values, so as to obtain the comprehensive quantitative evaluation results of the correctness test and to check whether the calculated results of the coupling theory and influence mechanism meet the index requirements.

7.2.2 Coincidence Degree

Coincidence degree is a measure of the closeness of the theoretically calculated and actually tested values with the consideration of the importance of each index. With the percentage difference between the theoretically calculated and measured values, a certain score is obtained based on certain rules, and the comprehensive evaluation of coincidence degree between the theoretically calculated and measured values is obtained after synthesis evaluation [7–9]. The formula for calculating the coincidence degree is

$$S_F = \sum_{i=1}^{n} \omega_i S_i \tag{7.21}$$

where S_F is the total evaluation score of coincidence degree between the theoretical calculated value and the measured value, ω_i is the weight of the ith electrical performance index in the overall electrical performance, i.e. the weighting value, which can be obtained objectively based on the experience of the industry experts [10, 11], and S_i is the single score of the ith electrical performance (single item), which is obtained from the fuzzy cognitive curve. As shown in Figure 7.3, it is assumed that the upper and lower limits of the percentage difference between the theoretically calculated and measured values of the ith electrical performance are $\bar{x}_i$ and $\underline{x}_i$, respectively; then, x_i is in the interval according to the size of its value ($0 \leq x_i \leq 1$). The individual scores are obtained on the corresponding Y-axis through the scoring curve $S\_E\_x_i$ as follows:

$$S\_E\_x_i = \begin{cases} 100 & x \leq \underline{x} \\ 100 \exp\left[-\left(\dfrac{x-\underline{x}}{\sigma}\right)^2\right], & x > \underline{x} \end{cases}$$

Figure 7.3 Single electrical performance
scoring curve.

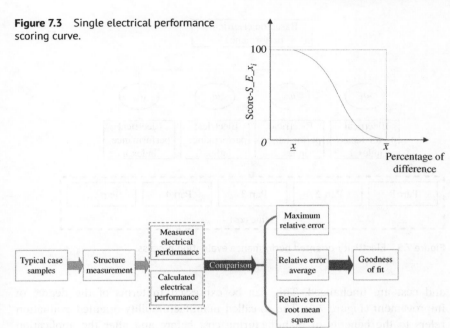

Figure 7.4 Flowchart of the calculation of the coincidence degree.

The specific values are different for different electrical properties of different cases. The calculating principle of the values of $\underline{x}$ and $\bar{x}$ is that it is assumed that when the ith percentage of difference between the theoretical calculated value and the measured value of electrical performance is 15%, $S\_E\_x_i = 60.0$. Note that the above calculation generally takes $\sigma = \frac{1}{3}(\bar{x} - \underline{x})$.

In the actual evaluation process, the maximum value, the average value, and the root mean square value of the relative errors are also obtained, and the coincidence degree will be calculated from the average value of the relative errors, as shown in Figure 7.4.

7.3 On Validation of EMC Theory and IM

The effectiveness evaluation is divided into two aspects: electrical performance oriented and manufacturability oriented. The electrical performance-oriented evaluation is to see how much the electrical performance is improved, i.e. the electrical performance improvements before and after applying the EMC theory and influence mechanism, while the original manufacturing process

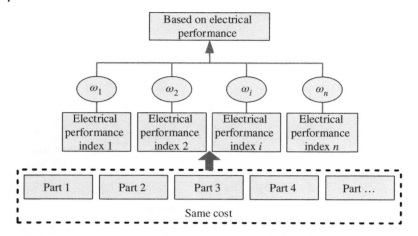

Figure 7.5 Electricity-oriented performance evaluation process.

and cost are unchanged. This can be expressed in terms of the degree of improvement (Figure 7.5). The so-called manufacturability-oriented evaluation refers to the benefits in manufacturing cost before and after the application of EMC theory, i.e. the cost reduction due to the relaxation of manufacturing accuracy and process requirements, while the electrical performance remains unchanged.

The degree of improvement is a measure of the electrical performance improvement after the application of EMC theory compared with the original design scheme. By considering the importance of each index, a certain score is obtained by certain rules, and the evaluation of the degree of comprehensive electrical performance improvement can also be obtained after synthesis evaluation [12, 13]. The enhancement degree is

$$S_P = \sum_{i=1}^{n} \omega_i \cdot y_i \tag{7.22}$$

where S_P is the overall electrical performance score, ω_i is the weighting value of the ith electrical property in the overall electrical properties, and y_i is the score of the ith electrical property in the overall electrical properties.

It is assumed that the ith value before and after the electrical performance improvement be set as P_i^0 and P_i^1; then, there are

$$x_i = P_i^1 - P_i^0 \tag{7.23}$$

Figure 7.6 shows the scoring curves corresponding to different types of performances, given a certain performance parameter x_i, whose individual scores can

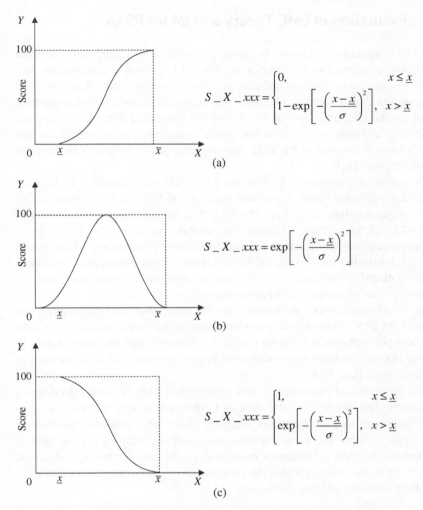

Figure 7.6 Electrical performance-based scoring curve. (a) Biased large type. (b) Intermediate type. (c) Biased small type.

be obtained on the corresponding *Y*-axis. The scoring curves can be obtained from objective experience by fitting curves, or described by fuzzy affiliation discriminant curves, which have specific mathematical expressions, so that the scores can be obtained easily. The scoring curves are different due to the different tendency for different electrical properties, which are (i) biased large type, (ii) intermediate type, and (iii) biased small type.

7.4   Evaluation of EMC Theory and IM for PSAA

The PSAA structure is shown in Figure 3.7, and its structure and electrical performance parameters are listed in Section 6.3.1. For the evaluation, the proposed correctness verification and effectiveness evaluation methods are specially integrated in the comprehensive measurement and evaluation prototype system. Based on the measured data of PSAA and the calculated data of the coupling theory and influence mechanism, the results of the correctness verification and effectiveness evaluation of the field coupling theory and influence mechanism can be obtained [14].

Three frequency points of 9.45, 9.60, and 9.75 GHz are considered. The electrical properties are mainly gain, horizontal and vertical first sidelobe levels (SLLs), and voltage standing wave ratio (VSWR). The specific results are shown in Tables 7.1–7.3, where Table 7.1 shows the maximum value, mean value, and root mean square value of the relative errors of electrical performances at 9.45, 9.60, and 9.75 GHz for the five sets of samples. Tables 7.2 and 7.3 show the effectiveness evaluation of electrical performance- and manufacturability-oriented, respectively.

The electrical performance improvement degree is 11.47.

The above correctness verification results show that the maximum error between the EMC theory calculation results and the measured results of the main electrical performance is no more than 9.9%, among which the error of gain in the key electrical performance parameters is no more than 2.3%, and the error of SLL is no more than 9.44%.

Both the electrical performance- and manufacturability-oriented effectiveness evaluation data show that the first SLL (horizontal and vertical) is averagely improved after applying the coupling theory and influence mechanism with approximately the same manufacturing cost significantly. In the case of unchanged electrical performance requirements, the manufacturing cost can be reduced by 27.04% after applying the coupling theory and influence mechanism, the effectiveness of which is easily seen.

7.5   Evaluation of EMC Theory and IM for a Radar Servo Mechanism

The structure of this airborne 3D antenna mount is shown in Figure 6.8, and the structural and servo system performance parameters and indexes are listed in Section 6.3.2. To perform the evaluation, the established correctness verification and effectiveness evaluation methods are specially integrated in a comprehensive measurement and evaluation prototype system, and the correctness verification

Table 7.1 Field coupling theory and influence mechanism correctness verification results.

| | | 9.45 GHz | | |
| | | | | |

Error \ Electrical performance	Horizontal maximum SLL	Vertical maximum SLL	Gain	VSWR
Maximum relative error	7.36%	3.99%	2.29%	8.97%
Average relative error	2.27%	1.792%	1.79%	7.708%
Relative error root man square	2.924%	1.628%	0.484%	2.099%

| | | 9.60 GHz | | |
| | | | | |

Error \ Electrical performance	Horizontal maximum SLL	Vertical maximum SLL	Gain	VSWR
Maximum relative error	3.92%	4.38%	1.88%	9.84%
Average relative error	2.032%	1.248%	1.226%	6.764%
Relative error root mean square	1.268%	1.756%	0.599%	2.546%

| | | 9.75 GHz | | |
| | | | | |

Error \ Electrical performance	Horizontal maximum SLL	Vertical maximum SLL	Gain	VSWR
Maximum relative error	9.44%	7.41%	1.66%	7.82%
Average relative error	5.59%	3.228%	1.178%	6.47%
Relative error root mean square	3.045%	2.205%	0.512%	1.215%

| | | Five samples | | |
| | | | | |

Error \ Electrical performance	Horizontal maximum SLL	Vertical maximum SLL	Gain	VSWR
Maximum relative error	9.44%	7.41%	2.29%	9.84%
Average relative error	3.297%	2.089%	1.398%	6.771%
Relative error root mean square	2.412%	1.963%	0.532%	1.953%
Agreement degree		90.17		

Table 7.2 Electrical performance-oriented effectiveness evaluation results.

9.45 GHz

Comparison / Index	Horizontal maximum SLL	Vertical maximum SLL	Gain	VSWR
Before application	−23.21	−25.47	37.35	1.8
After application	−27.11	−31.5	37. 37	1.8
Boost/lower percentage	16.8%	23.67%	0.05%	0.0%

9.6 GHz

Comparison / Index	Horizontal maximum SLL	Vertical maximum SLL	Gain	VSWR
Before application	−27.69 dB	−29.55 dB	37.64 dB	1.8
After application	−31.12 dB	−29.68 dB	37.72 dB	1.8
Boost/lower percentage	12.39%	0.44%	0.21%	0.0%

9.75 GHz

Comparison / Index	Horizontal maximum SLL	Vertical maximum SLL	Gain	VSWR
Before application	−32.47 dB	−29.72 dB	37.87 dB	1.8
After application	−30.05 dB	−26.70 dB	37.85 dB	1.8
Boost/lower percentage	−7.45%	−10.16%	−0.05%	0.0%

Table 7.3 Manufacturability-oriented (cost) evaluation results.

Typical cases / Cost	Cost before application (¥10 000)	Cost after application (¥10 000)	Cost reduction after application
PSAA	128.16	93.5	27.04%

and effectiveness evaluation results are derived from the measured data and the computational data of the coupling theory and influence mechanism [15].

Here, the performance of pointing accuracy, servo bandwidth, adjustment time, and overshoot of three axes such as pitch, azimuth, and roll are concerned, and the specific results are shown in Tables 7.4–7.6, where Table 7.4 provides the maximum, mean, and root mean square values of relative errors of servo performance

Table 7.4 Effect mechanism correctness verification results.

Azimuth axis

Error \ Electrical performance	Rise time (s)	Adjustment time (s)	Overshoot amount (%)	Pointing accuracy (mrad)	Servo bandwidth (Hz)
Maximum relative error	6.14%	8%	10.56%	10.71%	9.28%
Average relative error	3.196%	4.954%	6.732%	6.532%	7.422%
Relative error root mean square	2.625%	3.239%	2.778%	3.714%	1.701%

Pitch axis

Error \ Electrical performance	Rise time (s)	Adjustment time (s)	Overshoot amount (%)	Pointing accuracy (mrad)	Servo bandwidth (Hz)
Maximum relative error	8.75%	7.69%	9.17%	10.61%	9.35%
Average relative error	3.838%	4.954%	7.254%	7.28%	4.598%
Relative error root mean square	2.78%	3.219%	1.691%	2.872%	2.825%

Cross-rolling axis

Error \ Electrical performance	Rise time (s)	Adjustment time (s)	Overshoot amount (%)	Pointing accuracy (mrad)	Servo bandwidth (Hz)
Maximum relative error	8.25%	9.64%	5.63%	10.53%	7.42%
Average relative error	5.388%	4.508%	3.29%	7.82%	6.682%
Relative error root mean square	3.002%	3.523%	1.527%	2.989%	0.645%

Five samples

Error \ Electrical performance	Rise time (s)	Adjustment time (s)	Overshoot amount (%)	Pointing accuracy (mrad)	Servo bandwidth (Hz)
Maximum relative error	8.75%	9.64%	10.56%	10.71%	9.35%
Average relative error	4.141%	4.805%	5.759%	7.211%	6.234%
Relative error root mean square	2.802%	3.327%	1.999%	3.191%	1.724%
Agreement degree			82.35		

Table 7.5 Electrical performance-oriented effectiveness evaluation results.

Azimuth axis

Comparison ╲ Indicators	Rise time (s)	Adjustment time (s)	Overshoot amount (%)	Pointing accuracy (mrad)	Servo bandwidth (Hz)
Before application	0.066	0.1404	13.13	0.562	6.239
After application	0.04	0.06	2.41	0.15	8.65
Boost/lower percentage	39.12%	57.26%	81.64%	73.3%	38.64%

Pitch axis

Comparison ╲ Indicators	Rise time (s)	Adjustment time (s)	Overshoot amount (%)	Pointing accuracy (mrad)	Servo bandwidth (Hz)
Before application	0.06	0.1723	13.13	0.641	5.545
After application	0.04	0.09	2.15	0.21	8.55
Boost/lower percentage	33.33%	47.76%	83.62%	67.24%	54.19%

Cross-rolling axis

Comparison ╲ Indicators	Rise time (s)	Adjustment time (s)	Overshoot amount (%)	Pointing accuracy (mrad)	Servo bandwidth (Hz)
Before application	0.087	0.13	7.9	0.38	5.27
After application	0.085	0.09	2.1	0.25	5.8
Boost/lower percentage	2.3%	30.77%	73.42%	34.21%	10.06%

Table 7.6 Manufacturability-oriented (cost) evaluation results.

Typical cases ╲ Cost	Cost before application (¥10 000)	Cost after application (¥10 000)	Cost reduction after application
Three-dimensional antenna holders	20.2774	13.7668	32.1%

indexes for each axis of pitch, azimuth, and roll for five sample sets. Tables 7.5 and 7.6 show the evaluations for electrical performance and manufacturability, respectively.

The servo performance improvement degree is 72.29.

The results of the correctness verification data show that the maximum error between the calculated results and the actual test results of all servo performance parameters does not exceed 10.8%, among which the error of pointing accuracy in key servo performance does not exceed 10.8%, the error of adjustment time does not exceed 9.7%, and the error of rise time does not exceed 8.8%.

The evaluation results for electrical performance and manufacturability show that the performances of adjustment time, overshoot, and pointing accuracy are significantly improved after applying the coupling theory and influence mechanism with the same manufacturing cost approximately. In the case of unchanged electrical performance requirements, the manufacturing cost decreases by 32.1% after applying the coupling theory and influence mechanism, and the results are obvious.

7.6 Evaluation of EMC Theory and IM for Filter

The structure of the ESC duplex filter is shown in Figure 6.14, and the structural and electrical performance parameters and indexes are described in Section 6.3.3. For the evaluation, the established correctness verification and effectiveness evaluation methods are specially integrated in a comprehensive measurement and evaluation prototype system, and the results of the correctness verification and effectiveness evaluation of the field coupling theory and influence mechanism of this filter can be obtained based on the measured data and the computational data of the coupling theory and influence mechanism [16].

Here the low-, middle-, and high-frequency bands of the filter are concerned, and the frequency points are chosen as 610, 790, and 960 MHz, respectively. The electrical properties mainly consist of bandwidth, insertion loss, return loss, and stopband rejection. The results are shown in Tables 7.7–7.9, where Table 7.7 shows the maximum, mean, and RMS values of the relative errors of the electrical performance indexes for the five sets of samples at three frequency points of 610, 790, and 960 MHz. The electrical performance- and manufacturability-oriented evaluations are presented in Tables 7.8 and 7.9, respectively.

The electrical performance improvement degree is 64.3.

From the above results, it can be seen that the error between the theoretical calculation results and the actual test results of all electrical performances does not exceed 12.0%, among which the error of key electrical performance parameters

Table 7.7 ESC duplex filter correctness verification results.

610 MHz					
Electrical performance Error	Center frequency (MHz)	Return loss (dB)	Blocking band rejection (dB)	Bandwidth (MHz)	Insertion loss (dB)
Maximum relative error	0.16%	9.18%	0.78%	10%	11.94%
Average relative error	0.032%	5.54%	0.422%	2%	7.87%
Relative error root mean square	0.072%	2.853%	0.284%	4.472%	3.722%

790 MHz					
Electrical performance Error	Center frequency (MHz)	Return loss (dB)	Blocking band rejection (dB)	Bandwidth (MHz)	Insertion loss (dB)
Maximum relative error	0.06%	7.85%	0.94%	7.14%	7.79%
Average relative error	0.036%	4.628%	0.552%	3.572%	5.132%
Relative error root mean square	0.033%	3.312%	0.385%	2.526%	3.064%

960 MHz					
Electrical performance Error	Center frequency (MHz)	Return loss (dB)	Blocking band rejection (dB)	Bandwidth (MHz)	Insertion loss (dB)
Maximum relative error	0.26%	10.07%	7.69%	9.68%	6.88%
Average relative error	0.094%	6.478%	3.132%	3.248%	3.87%
Relative error root mean square	0.113%	3.275%	2.799%	3.952%	2.955%

Five samples					
Electrical performance Error	Center frequency (MHz)	Return loss (dB)	Blocking band rejection (dB)	Bandwidth (MHz)	Insertion loss (dB)
Maximum relative error	0.26%	10.07%	7.69%	10%	11.94%
Average relative error	0.054%	5.549%	1.369%	2.94%	5.624%
Relative error root mean square	0.073%	3.146%	1.156%	3.65%	3.067%
Agreement degree			87.926		

Table 7.8 Electrical performance-oriented effectiveness evaluation results.

610 MHz					
Comparison ⟍ Indexes	Center frequency (MHz)	Return loss (dB)	Blocking band rejection (dB)	Band width (MHz)	Insertion loss (dB)
Before application	611	−6.0564	−58.0172	20	−0.8949
After application	610	−12.094	−58.2352	20	−0.58496
Boost/lower percentage	5%	99.69%	0.38%	0.0%	34.63%

790 MHz					
Comparison ⟍ Indexes	Center frequency (MHz)	Return loss (dB)	Blocking band rejection (dB)	Band width (MHz)	Insertion loss (dB)
Before application	789	−7.3298	−46.0307	28	−0.4684
After application	790	−13.3938	−48.6003	30	−0.36484
Boost/lower percentage	5%	82.73%	5.58%	7.14%	22.11%

960 MHz					
Comparison ⟍ Indexes	Center frequency (MHz)	Return loss (dB)	Blocking band rejection (dB)	Band width (MHz)	Insertion loss (dB)
Before application	960.5	−7.5699	−43.9656	31	−0.8456
After application	960	−11.3174	−44.1296	34	−0.54855
Boost/lower percentage	2.5%	49.51%	0.37%	9.68%	35.13%

Table 7.9 Manufacturability-oriented evaluation results.

Typical cases ⟍ Cost	Cost before application (¥10 000)	Cost after application (¥10 000)	Cost reduction after application
ESC duplex filter	1.13951	0.87951	22.8%

does not exceed 12.0% for insertion loss, 10.1% for return loss, and 7.7% for stop-band rejection.

The numerical results for electrical performance and manufacturability show that the return loss and insertion loss performance improve significantly after applying the coupling theory and the influence mechanism with the same manufacturing cost approximately. In the case of constant electrical performance requirements, the manufacturing cost can be reduced by 22.8% after applying the coupling theory and influence mechanism, which is effective.

References

1 Dong, D. and Hua, Q. (2005). *Modern Comprehensive Evaluation Methods and Selected Cases*. Beijing: Tsinghua University Press (in Chinese).

2 William, B.R. (2003). Engineering complex systems: implications for research in systems engineering. *IEEE Transactions on Systems, Man, and Cybernetics—Part C: Applications and Reviews* 33 (2): 254–256.

3 Guohong, C., Yantai, C., and Meijuan, L. (2003). Research on the combination evaluation system. *Journal of Fudan University (Natural Science)* 42 (5): 667–672. (in Chinese).

4 Shang, J.S., Tjader, Y.X., and Ding, Y.Z. (2004). A unified framework for multi-criteria evaluation of transportation projects. *IEEE Transactions on Engineering Management* 51 (3): 300–313.

5 Girod, O.A. and Konstantinos, P.T. (1999). The evaluation of productive efficiency using a fuzzy mathematical programming approach: the case of the newspaper preprint insertion process. *IEEE Transactions on Engineering Management* 46 (4): 429–443.

6 Lin, C.T. and Chen, C.T. (2004). New product go/no-go evaluation at the front end: a fuzzy linguistic approach. *IEEE Transactions on Engineering Management* 51 (2): 197–207.

7 Wang, W. and Yuzhi, Z. (2006). Method of obtaining eigenvector for a fuzzy AHP. *International Journal of Control* 21 (2): 184–188. (in Chinese).

8 Wang Zeyan, G. and Hongfang, Y.X. (2003). A method of determining the linear combination weights based on entropy. *Journal of Systems Engineering-Theory & Practice* 23 (3): 112–116. (in Chinese).

9 Hongbao, M., Fengming, Z., and Kui, F. (2007). A method of combination weighting for multiple attribute decision making based on interval estimation. *Systems Engineering-Theory & Practice* 27 (6): 86–92. (in Chinese).

10 Yiqiang, W., Jinsheng, L., and Xuzhu, W. (1994). The concept of consistency and weights of judgment matrix in uncertain AHP. *Systems Engineering-Theory and Practice* 14 (4): 16–22. (in Chinese).

11 Liang and Guohua, W. (2002). Method of solving weights based on hierarchical and interactive model. *Journal of System Engineering* 17 (4): 358–363. (in Chinese).

12 Shoukang, Q. (2003). *Comprehensive Evaluation Principles and Applications*. Beijing: Electronic Industry Press (in Chinese).

13 Yonghui, H. and Sihui, H. (2000). *Comprehensive Evaluation Methodology*. Beijing: Science Press (in Chinese).

14 Zhijian, Y. (2009). Effect of mechanics structure on the electric performances of radiation element in planar slotted array antenna. *Telecommunication Engineering* 49 (6): 60–65. (in Chinese).

15 Gu, J.F., Ping, L.H., and Liu, J.C. (2008) Coupled vibration analysis for test sample resonating with vibration shaker. *The Third Asia International Symposium on Mechatronics (AISM2008)*, August 27-31, Sapporo, Japan.

16 Ma, H.B., Yang, D.W., and Zhou, J.Z. (2010). Improved coupling matrix extracting method for Chebyshev coaxial-cavity filter. *The Progress in Electromagnetic Research Symposium*, March 22–26, Xi'an, China.

8

EMC- and IM-based Optimum Design of Electronic Equipment

8.1 Introduction

Electronic equipment is a typical mechatronic device, and the structural and electromagnetic characteristics are mutually influenced and constrained. The purpose of the preceding electromechanical coupling modeling and analysis is still to design high-performance electronic equipment, which must start from the perspective of electromechanical coupling and disciplinary intersection [1]. Based on the field coupling model and influence mechanism, the theory and method of electromechanical coupling optimization design is proposed on the basis of electromechanical coupling theoretical model and influence mechanism. The following four aspects are discussed, respectively.

8.2 EMC- and IM-based Reflector Optimum Design

8.2.1 Mathematical Description of the Electromechanical Coupling Optimization Design

The electromechanical coupling optimization design of the reflector antenna can be mathematically described as follows:

$$\text{Find } \boldsymbol{\beta} = (\beta_1, \beta_2, \cdots, \beta_{N_d})$$

$$\text{Min } W \text{ or } -Gain \text{ and } E_{side} \tag{8.1}$$

$$\text{S.T. } Gain(\boldsymbol{\beta}) \geq G^0 \text{ and } E_{side} \leq E_{side}^0 \text{ or } W(\boldsymbol{\beta}) \leq W_{max} \tag{8.2}$$

$$\sigma_i(\boldsymbol{\beta}) \leq \sigma_{max}, \quad (i = 1, 2, \ldots, N_m) \tag{8.3}$$

$$\beta_{k\,min} \leq \beta_k \leq \beta_{k\,max}, \quad (k = 1, 2, \ldots, N_d) \tag{8.4}$$

Electromechanical Coupling Theory, Methodology and Applications for High-Performance Microwave Equipment,
First Edition. Baoyan Duan and Shuxin Zhang.
© 2023 The Institute of Electrical and Electronics Engineers, Inc. Published 2023 by John Wiley & Sons, Inc.

where β represents the structural (size, shape, topology, and type) design variables, W is the self-weight of the antenna, $Gain(\beta)$ and G^0 are the actual value of the antenna gain and the minimum allowable value, respectively, and E_{side} and E_{side}^0 are the actual value of the first sidelobe level and the allowable value, respectively. The calculation of the electrical performance of the reflector antenna is based on the two-field electromechanical coupling model discussed in Chapter 3, σ_i and σ_{max} are the actual and allowable values of the stress in the ith element, respectively, N_m is the total number of elements in the structural finite element model, and $\beta_{k\,min}$ and $\beta_{k\,max}$ are the lower and upper limits of the kth structural design variable, respectively.

Note that when applying the field coupling model with the influence mechanism, the random error on the surface of the reflector is determined in this way. That is, the mean and variance are obtained according to the specific processing process, and the panels on the same ring are considered to have the same mean and variance. The random error based on the mean and variance is then randomly generated by the computer, and the error is superimposed with the normalized systematic deformation error and then involved in the calculation of the electrical properties of the electromechanical field coupling model [2–4].

8.2.2 Numerical Simulation and Engineering Applications

In order to verify the correctness of the electromechanical field coupling model and its superiority compared with Ruze formula, it is applied to a 7.3 m diameter shipboard antenna and a 40 m diameter land antenna, and the reasonable and satisfactory results will be obtained.

Example 8.1 *A 7.3 m Diameter Shipboard Reflector Antenna (Figure 8.1)*
Parabolic reflector antennas are widely used in satellite telecommunication ground stations, ships, and vehicles, which are applied as receiving stations, receiving and switching stations. The antenna parameters are as follows: S/X band, focal diameter ratio 0.347, guaranteed accuracy and the preservation of strength wind speed 20 and 55 m/s, guaranteed strength wind speed 55 m/s, service environment temperature from −45 to 60 °C, and reflector surface accuracy requirement better than 0.5 mm (rms).

The antenna reflective surface is composed of a series of shells supported by a backup structure. The backup structure consists of three ring beams and 16 radiation beams, and the radiation beams are uniformly distributed around the central body. The reflective surface is made of aluminum plate and divided into 16 sub-panels. In the structural analysis, the reflective surface is treated as a shell element. To enhance the stiffness of the reflecting surface, Z-shaped aluminum beams are

Figure 8.1 7.3 m shipboard reflector antenna. Source: Sichuan Xingjie Technology Co., Ltd.

used as reinforcement. The finite element model of the whole antenna consists of 25 427 nodes, 4705 elements, 896 beam elements, and 3809 shell elements.

Six working conditions are considered:

1. Self-weighting when facing upward;
2. Self-weighting when facing upward and 20 m/s wind side blowing;
3. Self-weighting when facing upward and 30 m/s wind side blowing;
4. Self-weighting when horizontal;
5. Self-weighting when horizontal and 20 m/s wind side blowing;
6. Self-weighting when horizontal and 30 m/s wind side blowing.

Table 8.1 provides the calculation results of the gain loss coefficients derived from field coupling theory model, Ruze formula, and FEKO software based on the finite element deformation analysis, respectively. The antenna pointing error and gain loss are listed in Table 8.2.

From Table 8.1, it can be seen that the systematic error increases with increasing wind speed when the antennas face upward and are subjected to operating conditions 1, 2, and 3, resulting in a gradual reduction in the proportion of random error. The conclusion is that the difference in the gain reduction factor from the field coupling model and Ruze formula gradually increases from 0.0721% to 2.1521% and then to 9.5627%.

When the antenna is pointed horizontally and subjected to working conditions 4, 5, and 6, the systematic error increases with the increase in wind speed, resulting in a gradual reduction in the proportion of random error. As a result, the

Table 8.1 Gain loss coefficients of the 7.3 m parabolic reflector antenna.

Work conditions	FEKO results (%)	Field coupling model results (%)	Ruze formula results (%)
1. Self-weighting when facing upward	99.9839	99.9903	99.9183
2. Self-weighting when facing upward and 20 m/s wind side blowing	98.9441	99.4554	97.3601
3. Self-weighting when facing upward and 30 m/s wind side blowing	93.6204	97.5648	89.0493
4. Self-weighting when level	99.8895	99.9631	99.7533
5. Self-weighting when level and 20 m/s wind side blowing	94.2360	98.1153	89.8645
6. Self-weighting when level and 30 m/s wind side blowing	86.0814	90.8139	58.2770

Table 8.2 Pointing error and gain loss of the 7.3 m parabolic reflector antenna.

Work conditions	Pointing error ϕ_x (deg)	Pointing error ϕ_y (deg)	Gain loss (dB)
1. Self-weighting when facing upward	0.0255	−0.0007	−0.0004
2. Self-weighting when facing upward and 20 m/s wind side blowing	0.0284	0.1073	−0.0237
3. Self-weighting when facing upward and 30 m/s wind side blowing	0.0638	0.2414	−0.1071
4. Self-weighting when level	0.0001	−0.0001	−0.0016
5. Self-weighting when level and 20 m/s wind side blowing	0.0135	−0.0006	−0.0826
6. Self-weighting when level and 30 m/s wind side blowing	−0.0216	−0.0004	−0.4185

difference in the gain reduction factor from the field coupling model and Ruze formula gradually increases from 0.2103% to 9.1814% and then up to 55.8315%.

In addition, the gain reduction coefficients obtained by applying the field coupling model are higher than those by Ruze formula for the same systematic and random errors, and the former results are closer to the values obtained by full-wave FEKO software.

Example 8.2 *40 m Land-based Parabolic Reflector Antenna (Figure 8.2)*
The parameters of the 40 m parabolic reflector antenna are as follows: S/X band, focal diameter ratio of 0.33, guaranteed accuracy wind speed 20 m/s, guaranteed strength wind speed 40 m/s, service environment temperature from −10 to 50 °C, and reflector surface accuracy requirement better than 0.6 mm (rms).

The antenna reflector is a combined plate and net structure, of which the center part is plate, and the outer edge is net. The reflector is divided into 9 rings of 464 panels. The backup structure consists of a central body and a frame structure. Sixteen radial beams are evenly distributed around the central body. Six planar ring ribs are connected with crossbars. The finite element model of the structure includes 65 475 nodes and 143 748 elements, including 16 917 beam elements and 126 342 shell elements.

Under the same six working conditions as 7.3 m antenna, the antenna gain loss coefficients obtained by field coupling model, Ruze formula, and full-wave FEKO are shown in Table 8.3. The pointing error and gain loss are shown in Table 8.4.

As can be seen from Table 8.3, the systematic error increases with the increasing wind speed when the antennas face upward and are subjected to working conditions 1, 2, and 3, resulting in a gradual reduction in the proportion of random errors. The conclusion is that the difference in the gain reduction coefficient

Figure 8.2 40 m parabolic reflector antenna. Source: Wei WANG/Peiyuan LIAN/Shuxin ZHANG/Binbin XIANG/Qian XU/Springer Nature.

Table 8.3 Gain loss coefficients of 40 m parabolic reflector antenna.

Work conditions	FEKO results (%)	Field coupling model results (%)	Ruze formula results (%)
1. Self-weighting when facing upward	99.6321	99.8799	97.7304
2. Self-weighting when facing upward and 20 m/s wind side blowing	97.9510	99.4633	95.2294
3. Self-weighting when facing upward and 30 m/s wind side blowing	93.7996	97.5748	85.0986
4. Self-weighting when level	99.3769	99.7836	97.0041
5. Self-weighting when level and 20 m/s wind side blowing	95.8517	98.2508	87.8503
6. Self-weighting when level and 30 m/s wind side blowing	87.4024	92.2684	57.0322

Table 8.4 Pointing error and gain loss of 40 m parabolic reflector antenna.

Work conditions	Pointing error ϕ_x (deg)	Pointing error ϕ_y (deg)	Gain loss (dB)
1. Self-weighting when facing upward	0.0001	−0.0001	−0.0052
2. Self-weighting when facing upward and 20 m/s wind side blowing	−0.0072	−0.0288	−0.0234
3. Self-weighting when facing upward and 30 m/s wind side blowing	−0.0162	−0.0647	−0.1066
4. Self-weighting when level	0.0271	0.0001	−0.0094
5. Self-weighting when level and 20 m/s wind side blowing	0.0274	0.0001	−0.0767
6. Self-weighting when level and 30 m/s wind side blowing	0.0276	0.0001	−0.3495

derived from field coupling theory model and Ruze formula gradually increases from 2.1994% to 4.4460% and then to 14.6609%.

When the antenna is pointed horizontally and subjected to working conditions 4, 5, and 6, the systematic error increases with the increase in wind speed, resulting in a gradual reduction in the proportion of random error. As a result, the difference in the gain reduction coefficient by the field coupling model and Ruze formula gradually increases from 2.8653% to 11.8389% and then to 61.7830%.

In addition, the gain loss coefficient obtained by applying the field coupling model is higher than that by Ruze formula for the same systematic and random errors, and the former one is more close to the value obtained by applying the full-wave FEKO method.

Another point that should be noted is that the actual antenna in engineering will not have such high gain loss coefficient. The reason is that only the systematic and random errors of the main reflector are considered, because the purpose here is to compare field coupling theory model with the Ruze formula.

From the above results, it can be concluded that, first, for the same system with random errors, the electromechanical field coupling model yields a higher gain reduction coefficient than the Ruze formula because the field coupling theory model takes both errors into account, unlike the Ruze formula, which only considers random errors. In addition, the electromechanical coupling theory model takes into account the distribution of random errors on the reflector, unlike Ruze formula, which only takes into account a root mean square value when considering random errors. In turn, for the same electrical performance, the field coupling theory model requires lower surface error than the Ruze formula, which will lead to lower manufacturing cost. Obviously, it has important theoretical significance and engineering application value for the design and manufacturing of reflector antennas. Secondly, compared with Ruze formula, the more important advantage of the field coupling theory model is that the electrical properties are expressed as a function of structural design variables (such as size, shape, topology, and type of structure), which can realize the optimal structural stiffness distribution in the sense of electronic performance, thus making the electromechanical coupling optimization design be possible in theory.

8.3 EMC- and IM-based Cabinet Optimum Design

8.3.1 Mathematical Description of the Optimal Design of Electromechanical–Thermal Coupling

Based on the electromechanical–thermal field coupling theory model of the high-density chassis established in Chapter 3, the optimal design can be mathematically described as

$$\text{Find} \quad \beta = (\beta_1, \beta_2, \cdots, \beta_{N_d})^T$$

$$\text{Min} \quad W(\beta) \tag{8.5}$$

$$\text{S.T.} - SE(\boldsymbol{\beta}) + SE^0 \leq 0 \tag{8.6}$$

$$-f_{eigen} + f^0_{eigen} \leq 0 \tag{8.7}$$

$$T_{j\,\text{max}}(\boldsymbol{\beta}) - T^0_{j\,\text{max}} \leq 0 \quad (j = 1, 2, \cdots, N_u) \tag{8.8}$$

$$\sigma_i - [\sigma] \leq 0 \quad (i = 1, 2, \cdots, N_m) \tag{8.9}$$

$$\beta_{k\,\text{min}} \leq \beta_k \leq \beta_{k\,\text{max}} \quad (k = 1, 2, \cdots, N_d) \tag{8.10}$$

where the structural design variable $\boldsymbol{\beta} = (\beta_1, \beta_2, \cdots, \beta_{N_d})^T$ is the structural design parameter, which includes the length, width, and height of the chassis, the size and location of the heat sink slits, and the location of the internal devices; $W(\boldsymbol{\beta})$ is the weight of the structure; $SE(\boldsymbol{\beta})$ and SE^0 are the actual and allowable screen effects of the chassis, respectively; parameters of f_{eigen} and f^0_{eigen} are the actual value and the minimum allowable value of the chassis inherent frequency, respectively; parameters of $T_{j\,\text{max}}$ and $T^0_{j\,\text{max}}$ are the actual value of the temperature at the jth point and the maximum allowable value; σ_i and $[\sigma]$ are the actual and allowable values of stress of the ith element, respectively; $\beta_{k\text{min}}$ and $\beta_{k\text{max}}$ are the kth lower and upper limits of the variable, respectively; and N_d, N_u, and N_m are the total number of design variables, the total number of temperature constraints, and the total number of stress constraints, respectively.

In the above optimized design model, the optimization objectives and constraints can be transformed. For example, the electromagnetic screen effect is chosen as an objective, and the weight as a constraint. Multiobjective optimization can also be achieved by considering the indexes of several physical fields in combination with weighting coefficients [5–10].

8.3.2 Numerical Optimization Design of the Practical Chassis

The basic structure of the aluminum alloy chassis is shown in Figures 8.3 and 8.4, with length, width, and height of 575, 482, and 532 mm, respectively, and the material is aluminum alloy. The interior of the chassis consists of two parts, the upper part is installed with 12 PCBs and 2 power supplies, and the lower part is a cooling air duct with an inclined wind shield. The front panel of the chassis has two sets of cooling holes at the upper end and two fans at the lower end, three fans at the upper end of the rear panel, and two fans under the PCB board and above the wind shield. When the chassis works, the power supply module and PCB board devices heat up, the wind enters from the lower end of the front panel, and due to the role of the wind shield, the wind passes through the PCB board and power supply and then blows out from the upper end of the rear panel of the chassis, thus

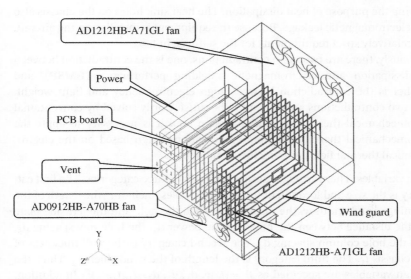

Figure 8.3 Schematic diagram of the internal components of the chassis for the simulation case.

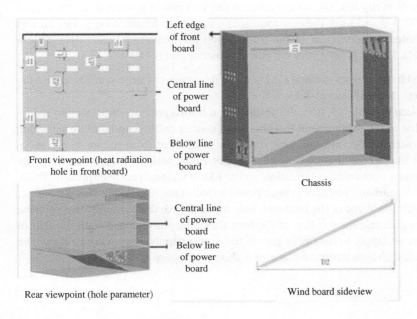

Figure 8.4 Chassis structure diagram.

achieving the purpose of heat dissipation. The heat sink holes on the chassis also cause electromagnetic leakage. Because the equipment is installed on the aircraft, it has relatively strict requirements for the self-weight too.

Obviously, there are two main contradictions, one is the contradiction between heat dissipation and electromagnetic shielding performance (**EMSP**), and the other is the contradiction between high eigenfrequency and light weight. These two contradictions are difficult to be satisfactorily solved by conventional electromechanical thermal separation design, and it is necessary to apply the electromechanical thermal coupling optimization design based on the electro-mechanical thermal field coupling theory model.

Design variables: There are eight design variables in three categories. The first category is the size and location of the heat sink holes, including the length of the openings l and width w, the distance between the hole and the left side plate d_1, the distance between the hole and the power d_2, the hole row spacing d_3, and the hole column spacing d_4. The second category is the wall thickness of the chassis t_1. The third category is the length of the wind shield t_2. Thus, the design variables are specified as $\beta = (l, w, d_1, d_2, d_3, d_4, t_1, t_2)^T$. In addition, letting $T^0_{j\,max} = 75\,°C$ ($j = 1, 2, ..., N_u$). The thermal analysis demonstrates that the highest temperature occurs at the two power sources, so it is only necessary to verify that the temperature at the two power sources does not exceed $75\,°C$. $SE^0 = 35$ dB, the electromagnetic waves are horizontally polarized and radiated perpendicular to the front panel, the frequency is 500 MHz. $f^0_{eigen} = 70$ Hz, $[\sigma] = 150$ MPa, and the power consumption is 160 W.

The Hooke–Jeeves method [11] is applied to solve the problem. The optimization results are shown in Table 8.5.

As seen in Table 8.5, first, the optimization results provide a 6.3% weight reduction due to the thinner wall thickness and shorter baffle length, while satisfying all constraints. Second, the heat sink hole is changed from square to rectangular, for the flat rectangular hole has less electromagnetic leakage than the square hole under horizontal polarization, so the EMSP is significantly improved. Third, the maximum temperature at both power supplies has decreased. This is because, although the shape of the heat sink hole has changed, the heat sink area remains unchanged, and the location of the heat sink hole has changed from the rear of the power supply to the upper part of the power supply, improving air flow. The optimized chassis structure diagram is shown in Figure 8.5.

Table 8.5 Numerical optimization results of the chassis.

Items	Name	Initial value	Optimization results	Lower limit	Upper limit
Design Variables	Chassis wall thickness t_1	4.5 mm	3.75 mm	2 mm	6 mm
	Length of the wind shield t_2	300 mm	50 mm	50 mm	400 mm
	Length of the opening l	10 mm	20 mm	5 mm	30 mm
	Width of the opening w	10 mm	5 mm	5 mm	50 mm
	Distance between the hole and the left side plate d_1	20 mm	50 mm	5 mm	80 mm
	Hole to power distance d_2	20 mm	20 mm	5 mm	80 mm
	Hole line spacing d_3	15 mm	15 mm	5 mm	50 mm
	Hole column spacing d_4	15 mm	15 mm	5 mm	30 mm
Objective	Chassis weight W	72.25/kg	67.71/kg	/	/
Constraints	Maximum stress	81.9/MPa	118/MPa		150/MPa
	Inherent frequency f	73.80/Hz	73.69/Hz	70/Hz	
	Electromagnetic screen effect SE	28.80/dB	42.07/dB	35/dB	
	Power 1 temperature T_1	71.9/°C	65.33/°C		75/°C
	Power 2 temperature T_2	73.98/°C	70.52/°C		75/°C

8.4 EMC- and IM-based Radar Servo Mechanism Optimum Design

Radar antenna servo system includes mechanical structure (antenna support structure) and the corresponding control system, both of which are closely related. The mechanical structure is not only the carrier of servo performance and often restricts the realization and improvement of the servo tracking performance. Unfortunately, the traditional design of the servo system has followed the path of separation of structure and control, that is the mechanical structure and control system are designed separately, and then through repeated iterations, tuning to achieve the required performance indicators. The drawback of this design method

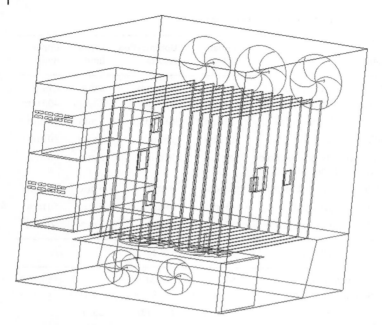

Figure 8.5 The optimized structure.

is that the structure design lacks in-depth analysis of the role of control, while the control design ignores the influence of structural factors, so the designed system is difficult to achieve the optimal overall performance.

With the increasing performance requirements on radar antenna servo system, the traditional design method of separating structure and control is becoming more and more ineffective, and it is necessary to integrate structure design and control design into a unified framework to pursue the overall optimal performance, which is the integrated design of structure and control.

8.4.1 Design Method of Structural Subsystem of Servo Control System

The purpose of structural design is to design a mechanical structure that meets the servo performance requirements. To obtain excellent servo tracking performance, the mechanical structure is generally required to be light weight and rigid, and these requirements are often contradictory. For this reason, structural optimization is introduced, i.e. the mass (inertia) distribution, transmission form, and topology are optimally designed to minimize the mass or volume while ensuring the stiffness requirements.

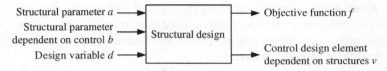

Figure 8.6 Structural optimization design of radar antenna servo system.

During the motion of the mechanism, its configuration is changing with time. For this reason, it can be formulated as a structural optimization problem PI as shown in Figure 8.6 with multiple operating conditions (which may be set to *nul* operating conditions).

PI: Find $d = (d_1, d_2, \cdots, d_{n_1})^T$

$$\text{Min} \quad f(a, b, d) = \sum_{i=1}^{n_2} W_i \rho_i \tag{8.11}$$

$$\text{S.T.} \quad -f_{re1}(d) = -\left[\left(\sum_{i=1}^{nul} f_{rei}^2\right) \bigg/ nul\right]^{\frac{1}{2}} \leq -\bar{f}_{re1} \tag{8.12}$$

$$\sigma_{ej} \leq [\sigma], (e = 1, \cdots, n_3; j = 1, \cdots, nul) \tag{8.13}$$

$$\delta_{ij} \leq \bar{\delta}_i, (i = 1, \cdots, n_4; j = 1, \cdots, nul) \tag{8.14}$$

Meanwhile, the following dynamic differential equation must be satisfied:

$$M_j \ddot{\delta}_j + E_j \dot{\delta}_j + K_j \delta_j = Z_j, (j = 1, \cdots, nul) \tag{8.15}$$

where n_1, n_2, n_3, and n_4 are the total numbers of design variables, components, stress constraints, and displacement constraints for the structural subsystem of the radar antenna servo system, respectively; symbols a and b are simple structural parameters (e.g. body dimensions and materials) and control-dependent structural design elements (e.g. driving forces), respectively. Equation (8.11) is the structure self-weight, W_i is the weight of the ith component, and ρ_i is the material density of the ith component. Equation (8.12) is the first-order intrinsic frequency constraint of the mechanism. The *nul* working condition is the *nul* typical positions of the mechanism, f_{rei} is the structural eigenfrequency at the ith typical position, $\bar{f}_{re1}$ is the minimum allowable value of the first-order inherent frequency of the mechanism. Equations (8.13) and (8.14) are the stress and displacement constraints, respectively, where σ_{ej} and $[\sigma]$ are the actual value and the maximum allowable value of the stress of the first element in the jth working condition, respectively. δ_{ij} and $\bar{\delta}_i$ are the actual and maximum allowable values of the ith displacement constraint at the jth working condition, respectively. Equation (8.15) is the dynamic differential equation to be satisfied by the structure in the first operating condition.

M_j, E_j, and K_j are the mass matrix, damping matrix, and stiffness matrix corresponding to the structure at the jth operating condition, respectively. $\ddot{\delta}_j$, $\dot{\delta}_j$, δ_j, and Z_j are the acceleration, velocity, displacement, and load array of the structure under the jth working condition, respectively.

Solving the nonlinear programming problems PI, the optimal values of the design variables and their corresponding structure-dependent control design elements are obtained (including M, E, K, and f_{rel}) as the basis for the control gain optimization design.

8.4.2 Design Method of the Control Subsystem of Servo System

The purpose of the control subsystem design is to design a control system that meets the performance requirements with the given structure. In general, the system has the requirements of stability, speed, and accuracy, i.e. the designed controller should achieve fast and accurate tracking of the target and ensure stability. To solve this, a control optimization design approach can be introduced, i.e. the controller gain p is optimally designed so that the system has excellent servo tracking performance and obtains control-dependent structural design elements B (e.g. drive force). The problem can then be described as a nonlinear programming problem PII, as shown in Figure 8.7.

PII: Find $\quad p = (p_1, p_2, \cdots, p_{NUP})^T$

$$\text{Min} \quad J(u, V, Z(p)) = \int_0^{T_0} e^2(t)dt \tag{8.16}$$

$$\text{S.T.} \quad Re[pole_i] < 0, (i = 1, \cdots NUI) \tag{8.17}$$

$$t_s \leq t_s^+ \tag{8.18}$$

$$\varsigma \leq \varsigma_{\max} \tag{8.19}$$

$$Z \leq Z_{\max} \tag{8.20}$$

$$Y(t) = \phi(t)e(t) \tag{8.21}$$

$$Z(t) = Z(e(t), p_i) \tag{8.22}$$

$$e(t) = Y(t) - Y_d(t) \tag{8.23}$$

$$\dot{V}(t) \leq 0 \tag{8.24}$$

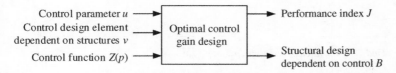

Figure 8.7 Optimal control gain design of the radar antenna servo system.

where p_i is the ith control gain variable, NUP is the total number of gain design variables, T_0 is a motion cycle, $e(t)$ and t_s are the tracking error and adjustment time, respectively, and ς is the overshoot amount. Obviously, the objective function J reflects the requirement for "fast" and "accurate" tracking performance. Equation (8.17) is the stability constraint of the system, and NUI is the total number of poles of the system. $\phi(t)$ can be understood as a "transfer function" that reflects the relationship between input $Y_d(t)$ and output $Y(t)$ in the time domain; $Z(t)$ is the driving force or torque of the controller in the time domain; $Y_d(t)$ is the control target; and $\dot{V}(t)$ is the derivative of the constructed Lyapunov function $V(t)$, and the goal is to ensure the stability of the controller, where

$$V(t) = \frac{1}{2}f_1\left(M_1, \dot{\theta}_1^2\right) + \frac{1}{2}f_2\left(M_2, \dot{\theta}_2^2\right) + \frac{1}{2}f_3(M_3, V^2) \tag{8.25}$$

8.4.3 Design Method with Integration of Structural and Control Technologies for Radar Servo System

For high-performance radar servo systems, it is difficult to achieve the required performance specifications even if the structure and control are optimized separately. Because the abovementioned methods do not guarantee that the designed servo system is optimal as whole, the possible result is that when the control design is carried out based on the results of the structure optimization design, it is difficult to obtain a solution that satisfies the performance index or obtain design elements that contradict with the structural optimization design. For this reason, it is necessary to carry out the integrated structural and control optimization design, that is integrate the structural optimization and control optimization. Specifically, the optimal values of the structural-design variable $\boldsymbol{d}$ and the control gain variable $\boldsymbol{p}$ are found by seeking the optimal integrated performance index [12–14] $\boldsymbol{H}$ under a given structural parameter a and control parameter u. Thereby, the problem can be described as a nonlinear programming problem PIII, as shown in Figure 8.8,

PIII

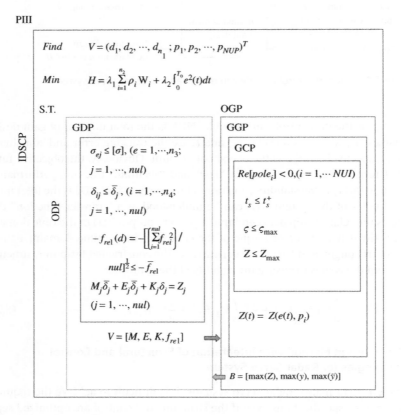

where GDP is the general structural design problem, which requires to satisfy the constraints on stresses, displacements, inherent frequencies, and dynamic differential equations. ODP is the optimal structural design problem, which requires to find the optimal structural design while satisfying the GDP constraint. Similarly, GCP is the general control problem, which requires nonlinear constraints such as stability, adjustment time, overshoot amount, and driving force (moment); GGP is the general control gain problem, which finds the driving force (moment) for structural design by selecting the control gain factor $\boldsymbol{p}$ while satisfying the GCP constraint. The OGP is the optimal control gain problem, i.e. the optimal selection of preferred control gain factor to seek the goal of tracking "fast" and "accurate" while satisfying the GCP and GGP constraints, and the IDSCP is the coupling design problem of structure and control, i.e. the optimal design problem of structure and control while satisfying the GCP and GGP constraints. The IDSCP is a coupled structural–control design problem, i.e. to achieve the goals

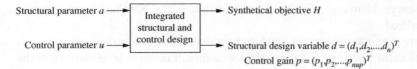

Structural parameter a ⟶ Integrated structural and control design ⟶ Synthetical objective H

Control parameter u ⟶ ⟶ Structural design variable $d = (d_1, d_2, ..., d_n)^T$
Control gain $p = (p_1, p_2, ..., p_{nup})^T$

Figure 8.8 Optimal design of the integrated structural and control of the radar antenna servo system.

of lightweight structure and stable, accurate, and fast control by optimizing the selection of structural variables and preferred control gain factor simultaneously while satisfying all constraints in the structural optimization ODP and control optimization OGP. V and B are the information interaction parts in the integrated optimization model. V is the structure-based control design element, which is different from the control parameter p, and B is the control-dependent structural design element, which is different from the structural design variable d.

The above model consists of two parts, namely, the optimal structural design problem (ODP) and the optimal control gain problem (OGP). The ODP internally includes the general design problem GDP, while the OGP includes the general gain problem (GGP) and the general control problem (GCP). The structural and control designs are performed through the link of structural-based control design elements $V = \begin{bmatrix} M & E & K & f_{rel} \end{bmatrix}$ and control-based structural design elements $B = \begin{bmatrix} \max(Z) & \max(\dot{y}) & \max(\ddot{y}) \end{bmatrix}$. In the integrated optimization, the two-objective function of the structural self-weight and the reflecting control performance is transformed into a single objective function by weighting factors, which are determined by the designer based on the importance of each single objective and satisfies $\lambda_i \geq 0 (i = 1, 2)$ and $\sum_{i=1}^{2} \lambda_i = 1$.

It is difficult because both the objective function and constraint function in the optimization model are high-order nonlinear functions for the design variables, and the analysis and design of the multiflexible system are also required. In addition, if the structure is variable, the transfer function method is no longer applicable, and how to deal with the stability constraints of the control is also a difficult problem. As can be seen, problem PIII is a very complex and highly nonlinear programming problem, which is very difficult to solve directly. For this reason, the objective function and the constraint function (except the dynamic differential equation and the stability constraint) are extended into second- and first-order Taylor series, respectively, so that it is transformed into a sequential quadratic programming (SQP) problem. The Lemke method is employed to solve the problem.

Problem PIII contains two types of design variables, structural factors and control gain factors, which have different dimensions and amplitudes, which may lead to illness problem and affect the iteration convergence. For this reason, it can be normalized.

There is another issue that needs to be noted, that is the variant structure versus invariant structure. For the variant structure, it needs to be performed in the time domain. In this case, the mechanism is partially transformed into a multicase structural problem, and the dynamic differential equations with stability constraints are used for calibration, without entering the optimization iteration. For the invariant structure, it can be carried out in the frequency domain, and the transfer function approach can be used. At this time, the mechanism is naturally a structural problem with a single working condition, and the dynamic differential equations and stability analysis are used only for calibration, and does not need to be introduced into the optimization iteration.

8.4.4 Numerical Simulation and Experimental Validation

In order to verify the feasibility and effectiveness of the method proposed in this book, the method is applied to several typical examples, and satisfactory results are obtained. Considering the limitation of the context, three examples are provided as follows: the first one is the numerical simulation result, and the other two are both numerical simulation and experimental verification.

Example 8.3 *Crank Slider Mechanism-type Reflector Antenna (Figure 8.9)*
A control torque is applied to the crank OA as shown in Figure 8.9, and a point α is selected on the connecting rod AB to mount the antenna, and $\alpha\beta$ corresponds to its pointing. The purpose is to make the antenna track the target accurately by adjusting the control torque and the structural design. The variation range of θ is from 10° to 80°.

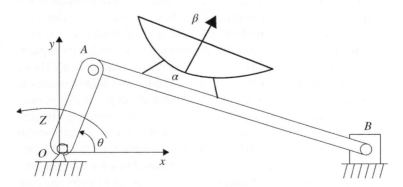

Figure 8.9 A crank linkage mechanism-type reflector antenna.

The crank and the connecting rod are hollow circular tubes, r_1, r_2 are the cross-sectional diameters of the crank and the connecting rod, respectively, and w_1, w_2 are the wall thickness. The proportional, integral, and differential gain parameters of the PID controller are p_1, p_2, and p_3. In the dynamics modeling, the crank is considered as a rigid body, and the connecting rod as a flexible body, and its elastic deformation is the superposition of the first n_e orders of the vibration shape of the simply supported beam, which is taken as $n_e = 3$ in this case.

In the optimization, $\lambda_1 = \lambda_2 = 0.5$, $Z_{max} = 100$ N. m, $t_s^+ = 0.15$ seconds, $\varsigma_{max} = 10\%$, $\bar{f}_{rel} = 10$ Hz, $[\sigma] = 150$ Mpa, the typical number of working conditions NUL is taken as 2, and $\theta_1 = 40°$, $\theta_2 = 70°$.

The results of both the coupling and the separation design are shown in Table 8.6. Figures 8.10 and 8.11 show the comparison curves of the response and driving torque for the first 0.2 seconds, respectively. There is a little difference between the two methods after 0.2 seconds. It can be seen that the results of the coupling and separation design are significantly better than those of the

Table 8.6 Comparison of coupling and separation design results of crank-link mechanism-type reflector antenna.

	Parameters	Lower bound	Upper bound	Initial value	Optimal values of separation design	Optimal values of coupling design
Design variables	r_1(mm)	5	15	8	14.958	7.407
	r_2(mm)	2	8	4	2	3.175
	w_1(mm)	1	6	2	1	1
	w_2(mm)	0.5	3	2	0.5	0.5
	p_1	0	60	22	48.3182	58.7571
	p_2	0	60	15	0	14.5971
	p_3	0	15	2	0.8362	0.9245
Indicators	t_s(s)	0	0.15	0.064	0.074	0.064
	ς	0	10%	5.85%	10%	9.16%
	σ_{max}(MPa)	0	150	23	150	147
	f_{rel}(Hz)	10		23.56	11.5189	18.2
Objective function	Weight (kg)			0.627313	0.268933	0.1867
	J(rad$^2 \cdot$ s)			0.073244	0.019872	0.0163

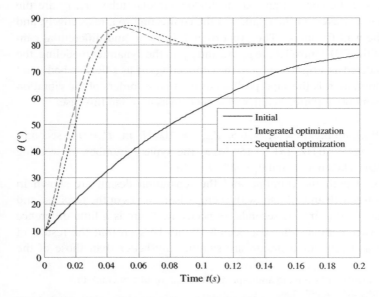

Figure 8.10 Comparison of motion simulation for coupling and separation designs.

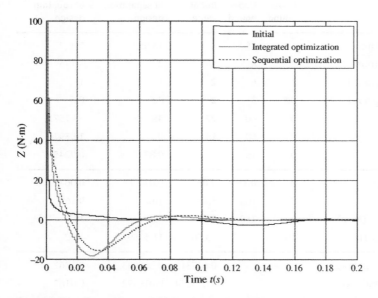

Figure 8.11 Comparison of driving torque for coupling and separation designs.

separated optimization, such as a 13.5% reduction in the tuning time (from 0.074 to 0.064 seconds), a 58.12% increase in the intrinsic frequency (from 11.51 to 18.2 Hz), and a 30.57% decrease in the total mass (from 0.2689 to 0.1867 kg).

Example 8.4 *Servo Lab Bench*

Consider a servo system consisting of a gear reducer as shown in Figure 8.12a and simplify it to a two-axis, three-inertia system as shown in Figure 8.12b. It is assumed that the equivalent rotational inertia on the motor shaft are J_1, J_2, and J_3, the torsional stiffness of the corresponding shafts are k_1 and k_2, the damping coefficients are b_1 and b_2, and the friction coefficients at the three bearings are d_1, d_2, and d_3, respectively.

When the load and motor are defined, the distance between the motor shaft and the load shaft is determined by the geometrical parameters, and the controller adopts the traditional digital PID, which requires the design of the corresponding structure and control parameters (including the load shaft length L, radius R, active shaft radius r, reduction ratio i, and PID control gain p_1, p_2, and p_3), to make the system achieve the overall optimal performance while meeting the required performance specifications (overshoot amount at unit step response $\varsigma_{max} \leq 2\%$ and adjustment time $t_s \leq 0.3$ seconds).

When the same initial values (initial design of the servo test bench) are used, both the separated optimization design and the coupled optimization design are carried out, respectively, and the sequential quadratic programming method is applied to solve the results as shown in Table 8.7, and the unit step response of the corresponding system is shown in Figure 8.13.

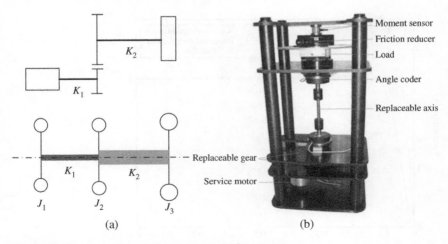

Figure 8.12 A servo lab bench. (a) Model. (b) Prototype.

Table 8.7 Comparison of the results of the coupled and separated optimization designs for a servo experimental bench.

Parameters		Lower bound	Upper bound	Initial value	Separated optimization	Coupled optimization
Design variables	R(m)	0.02	0.005	0.010	0.008	0.009
	L(m)	0.05	0.01	0.027	0.010	0.010
	i	12.0	2.00	6.625	6.228	11.950
	r(m)	0.01	0.002	0.005	0.003	0.002
	k_p	2.0	0.01	0.102	0.605	0.437
	k_i	0.01	0.0001	0.001	0.007	0.002
	K_d	1.2	0.001	0.082	0.416	0.198
Indicators	τ_{max} (N.m)	1.0	−1.0	0.102	1.000	1.000
	t_s (s)	0.3	0.0	0.23	0.097	0.07
	ς (%)	2.0	0.0	3.50	2.001	1.999
	σ_{max} (MPa)	30.0	0.0	0.2546	30.001	19.760
	f (Hz)	—	12.0	18.68	12.000	12.001
Objective function	Err(rad²s)	—			0.0577	0.0158	0.0103
	Weight (kg)	—			8.8279	7.9586	8.8012

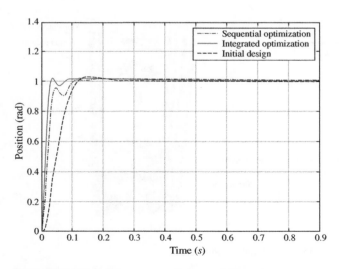

Figure 8.13 Unit step response diagram of the system.

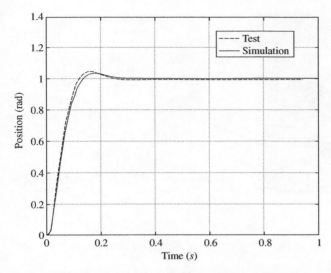

Figure 8.14 Simulation and experiment of the unit step response (initial design) of the system.

As can be seen from Table 8.7, the cumulative tracking error of the coupling design is reduced by 34.8% (from 0.0158 to 0.0103 rad^2·seconds) compared to the separated designs, while the weight only increases by 10.59% (from 7.96 to 8.80 kg), indicating that the overall performance of the coupling design is better than the separated design.

To illustrate the correctness of the results, the numerical results with the initial parameters were physically verified on the experimental bench. Figure 8.14 shows the comparison between the measured unit step response and the simulation results when the initial design is used. Since the dynamic characteristics and manufacturing accuracy of the motor and servo amplifier are not considered, the experimental results are a little different from the simulated results (maximum error < 5%). It should be noted that in order to do experiments for the optimized results, special custom-made gears, shafts, and the corresponding structures are required, which is not very practical. However, the experiments with initial parameters illustrate the accuracy of the established model.

Example 8.5 *A 40 m Antenna Servo System*
The azimuth rotating structure system of 40 m antenna is shown in Figure 8.15. The antenna reflector is mounted on the fork arm through the support base. The driving torque is generated by the azimuth servo motor and acts on the turntable through the reducer, drive shaft, and gear ring, which drives the antenna reflector to rotate around the azimuth axis. The antenna reflector weighs 65 tons, and its

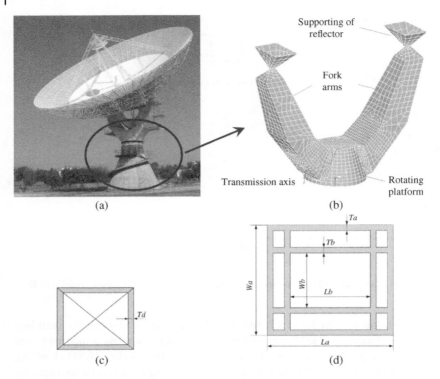

Figure 8.15 The azimuthal rotating body structure of 40 m parabolic reflector antenna. (a) Overall antenna structure diagram. (b) Fork arm support structure. (c) Schematic diagram of the cross-sectional structure of the antenna reflector support seat. (d) Schematic diagram of cross-sectional structure of fork arm.

tracking accuracy is required to be 30 arc seconds. It is assumed that the fork arm structure, the external dimensions, and the reduction ratio of the azimuth rotation system have already been determined. The purpose of the optimization design is to improve the tracking performance of the antenna and reduce the weight of the azimuth slewing system by adjusting the control torque and structural design. The structural design variables include the outer ring wall thickness of the upper and lower support arm box structures (T_a), the inner ring wall thickness of the upper and lower support arm box structures (T_b), the wall thickness of the rotary table structure (T_c), the wall thickness of the antenna support base (T_d), and the drive shaft radius (R). The control design variable is the PID gain coefficient (p_1, p_2, p_3).

In the optimization, the parameters are set as $Z_{max} = 18\,000$ N. m, $t_s^+ = 2.0$ seconds, $\varsigma_{max} = 2\%$, $\overline{f}_{rel} = 5$ Hz, and $[\sigma] = 30$ Mpa. The results of the optimal design of the coupled and separated optimization are shown in

Table 8.8 Optimal design results of the coupled and separated designs of 40 m antenna azimuth slewing system.

Parameters		Lower bound	Upper bound	Initial value	Separated optimization	Coupled optimization
Design Variables	Turntable wall thickness T_c (m)	0.20	0.01	0.050	0.0121	0.0151
	Wall thickness of the outer ring of the fork arm T_a (m)	0.20	0.01	0.050	0.0113	0.0112
	Wall thickness of the inner ring of the fork arm T_b (m)	0.20	0.01	0.030	0.0136	0.0133
	Wall thickness of the antenna support pedestal T_d (m)	0.20	0.01	0.030	0.0125	0.0122
	Drive shaft radius R (m)	0.01	0.002	0.005	0.0028	0.0034
	p_1	0.3	0.01	0.0312	0.0373	0.0436
	p_2	0.003	0.00	0.00001	0.00001	0.0000
	p_3	0.3	0.01	0.0415	0.0592	0.0695
Indicators	t_s (s)	2.0	0.0	5.31	1.890	1.620
	ς (%)	2.0	0.0	4.1	2.000	2.000
	σ_{max} (MPa)	30.0	0.0	12.971	29.999	30.000
	f_{rel} (Hz)		5.0	9.058	6.870	8.407
Objective function	Weight (T)			158.11	77.905	78.239
	$J(\deg^2 \cdot s)$				0.0042	0.0031	0.0026

Table 8.8. Figure 8.16 shows the comparison curves of the step response, and Table 8.8 shows the optimized design results. It can be seen that the adjustment time t_s is reduced by 14.3% (from 1.89 to 1.62 seconds) by the coupled optimization design, inherent frequency f_{rel} is increased by 22.42% (from 6.87 to 8.41), the cumulative tracking error is reduced by 16.12% (from 0.0031 to 0.0026), and the total mass is increased by 0.42% (from 77.91 to 78.24 tons). Clearly, the results of the coupling optimization design are obviously better than the results of the separated optimization design.

The above results illustrate that it is difficult and sometimes impossible to obtain the optimal overall performance by the separated design of structure and control, and the coupled optimization design can effectively solve this problem. The coupling optimization design is especially suitable for the scheme design of antenna servo systems.

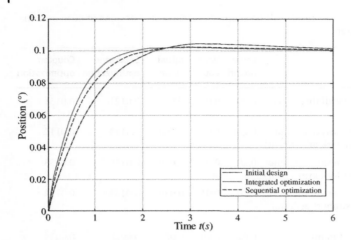

Figure 8.16 The step response curve of the 40 m antenna base azimuth axis.

8.5 A General Design-vector-based Optimum Design of Electronic Equipment

The optimal design of multiphysics field coupling problems is based on the mathematical model of CMFP with the solution strategies and methods. Several typical electromechanical coupling optimization design problems for electronic equipment are discussed above and partially applied. However, a higher level should be performed in the establishment of the optimal design model of the multiphysical field coupling system based on the unified design vector, which is described as follows.

The above idea can be implemented using a modular approach [15]. On the basis of the optimization and field analysis modules, the optimization model and strategy for the electromechanical–thermal coupling problem is realized by establishing the linkage between the modules through the design element approach [16, 17].

To solve this, a unified design vector is introduced $X = (x_1, x_2, \ldots, x_{nus})^T$, and the overall optimization model of the electromechanical–thermal coupling problem is established as follows:

Find $X = (x_1, x_2, \ldots, x_{nus})^T$

Min $z(X), X \in R^{n_x}$ $\qquad\qquad$ (8.26)

$$\text{S.T. } g_i(X) \le 0, g_i \in R^{n_g}, (i = 1, 2, \ldots, m) \tag{8.27}$$

$$h_j(X) \le 0, h_j \in R^{n_h}, (j = 1, 2, \ldots, n) \tag{8.28}$$

$$X_L \le X \le X^U \tag{8.29}$$

where $z(X)$ is the objective function; $g_i(X)$ and $h_j(X)$ are nonlinear inequality and equality constraints; and X_L, X_U are the lower and upper bounds of the uniform design vector.

The uniform design vector X is a collection of design parameters in each physical field. In order to reduce the number of unified design vectors and improve the efficiency of optimization analysis, the introduced design elements will be organically linked with the design parameters of each physical field.

Secondly, the field analysis module adopts a kind of five-field coupled analysis model; in addition to the analysis models of structure, electromagnetic, and thermal, the motion information model of electromechanical and mechanical–thermal grids is added, namely

$$S(X, U, X_e, X_t) = 0 \quad \text{Structural displacement field} \tag{8.30}$$

$$E(X, V, X_e) = 0 \quad \text{Electromagnetic field} \tag{8.31}$$

$$R(X, U, X_e) = 0 \quad \text{Structural and electrical mesh} \tag{8.32}$$

$$T(X, W, X_t) = 0 \quad \text{Thermal field mesh} \tag{8.33}$$

$$D(X, U, X_t) = 0 \quad \text{Structural and thermal mesh} \tag{8.34}$$

where U, V, and W are the displacement vectors of the structural displacement field, the electromagnetic vectors of the electromagnetic field, and the temperature vectors of the thermal field, respectively; X_e, X_t are the mesh displacement vectors for electromagnetic field and thermal field, respectively; and Eqs. (8.32) and (8.34) describe the transformation of information between the structural displacement field and the electromagnetic field and the structural displacement field and the thermal field, respectively. Note that the design vector X are provided to the optimization model.

At this point, the overall framework of the optimal design for the electromechanical–thermal coupling problem of electronic equipment can be obtained. Of course, it is necessary to consider the solution method of the optimization model, the implementation method of the design element, the solution method of the coupling model, and the derivation method of the sensitivity equation. This needs a further deep and systematic research work.

References

1 Duan, B.Y. (1998). *Antenna Structure Analysis, Optimization and Measurement*, 115–138. Xi'an: Xi'an University of Electronic Science and Technology Press.

2 Wang, C.S., Duan, B.Y., and Qiu, Y.Y. (2007). On distorted surface analysis and multidisciplinary structural optimization of large reflector antennas. *International Journal of Structural and Multidisciplinary Optimization* 33 (6): 519–528.

3 Wang, C.S., Duan, B.Y., and Fei, Z. (2008). Structural electromagnetic-mechanical optimization design of large space truss plane antenna. *Journal of Electronics* 36 (9): 1776–1781.

4 Ma, H.B., Duan, B.Y., and Wang, C.S. (2009). Deformed reflector antenna with random factors and integrated design with mechanical and electronic syntheses. *Chinese Journal of Radio Science* 24 (6): 1065–1070.

5 Duan B Y, Wang C S. Analysis and optimization design of multi-field coupling problem in electronic equipment. *International Workshop 2007: Advancements in Design Optimization of Materials, Structures and Mechanical Systems, 17–20 December 2007*, Xi'an, China, 252–261.

6 Duan, B.Y. (2010). The multi-field-coupled model and optimization of absorbing material's position and size of electronic equipment. *Journal of Mechatronics and Applications* 1–6.

7 Guoqiang, S. (2009). Design of electronic cabinet reinforcement based on three-field coupling. Xidian University.

8 Yue, H. (2009). Thermal design of electronic device layout based on three-field coupling. Xidian University.

9 Hui, Q. (2009). Optimal design of absorbing materials for electronic devices based on three-field coupling. Xidian University.

10 Shibo, J. (2009). Study of three-field coupling analysis of electronic devices. Xidian University.

11 Fletcher, R. (1970). A new approach to variable metric algorithms. *Computer Journal* 13: 317–322.

12 Sulan, L., Jin, H., and Duan, B.Y. (2010). Integrated design of structure and control for Radar antenna servo-mechanism. *Journal of Mechanical Engineering* 46 (1): 140–146.

13 Hong, B., Duan, B.Y., and Jingli, D. (2008). Simultaneous optimization of control and structural systems of a complex machine. *Chinese Journal of Computational Mechanics.* 25 (1): 8–13.

14 Huang, J. (2009). HLA based multidisciplinary joint simulation technology for servo mechanism analysis. *IEEE International Conference on Mechatronics and Automation* 2009 (8): 4517–4522.

15 Deb, K. (1998). *Optimization for Engineering Design: Algorithms and Examples*, 105–136. New Delhi: Prentice-Hall of India.

16 Duan, B.Y. and Ye, S.H. (1986). A mixed method for shape optimization of skeletal structures. *International Journal of Engineering Optimization* 10 (3): 183–197.

17 Raulli M and Maute K (2002). Symbolic geometric modeling and parameterization for multiphysics shape optimization. In: *9th AIAA/ISSMO Conference on Multidisciplinary Analysis and Optimization*. Atlanta, USA, 1–10, IAA 2002-5648.

9

Computer Software Platform for Coupling Analysis and Design of Electronic Equipment

9.1 Introduction

In order to transform the field coupling theory model, influence mechanism, and testing and evaluation methods proposed above into tools that can be applied in engineering, it is necessary to develop an engineered professional industry software platform. This software can provide auxiliary software tools for the electromechanical coupling problems that are difficult to comprehensively analyze and design due to the original disconnection and then realize the cross-disciplinary coupling analysis and design of electronic equipment. This chapter focuses on the comprehensive platform system for electromechanical coupling analysis and design of electronic equipment, including the electromechanical–thermal coupling theory, the influence mechanism of structural factors on electrical performance, and the prototype software system for testing and evaluation.

9.2 General Method and System Project

The electromechanical coupling of electronic equipment involves the relationship among electromagnetic, mechanical, temperature, and other physical fields and physical quantities. It is highly professional and has many influencing factors. The analysis process of electromechanical coupling will be cyclical, and whether it is handled properly or not will directly affect the quality of the interactive interface of the comprehensive platform system and the level of calculation efficiency. First, it is necessary to analyze professional software such as structural analysis, thermal analysis, electromagnetic analysis, and geometrical modeling and perform coupling analysis in combination with specific examples to discover the characteristics of the existing professional software as well as the existent problems of input and output interfaces, model conversion, etc. After deep research and repeated

Electromechanical Coupling Theory, Methodology and Applications for High-Performance Microwave Equipment, First Edition. Baoyan Duan and Shuxin Zhang.

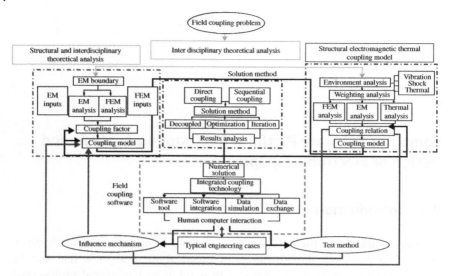

Figure 9.1 Flow chart of EM–structure–temperature coupling analysis.

attempts, the effective implementation of structural deformation, electromagnetic and temperature fields coupling analysis process is finally presented as shown in Figure 9.1. First of all, it is necessary to integrate the software and hardware resources and reasonably carry out system planning, resource allocation, problem estimation, task scheduling, etc. Through the actual analysis of typical objects, we can understand the principles of determining the scale of grid division, the criteria for determining the convergence of numerical calculations, the configuration and application of computer resources, and other issues. Furthermore, it can be used as the default parameters to provide guidance for the analysis and solution of the electromechanical coupling problem of electronic equipment. Secondly, in the process of modeling and solving electromechanical coupling problem, a comprehensive platform system for numerical analysis of electromechanical coupling characteristics is gradually constructed combining with the development of typical electronic equipment.

9.3 Integration of the Professional Software

On the basis of relevant professional software, the procedure is making full use of the existing mature advanced technology, focusing on the development of solution strategies and methods for electromechanical coupling problems, comprehensive software subsystems, and comprehensive design platforms. According to the overall and analysis process of the software design, the relevant professional software

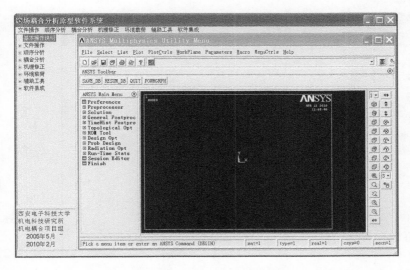

Figure 9.2 Comprehensive integration of structural analysis software.

is integrated into the coupling analysis platform, and it is used as a module that can be called at any time according to the analysis process to complete the comprehensive integration from content to detailed form.

The specific processing method is to set the path of the professional software, call the professional software internally, obtain the process of the professional software, control the size of its window, and place it in the designated subwindow of the electromechanical coupling analysis software. The professional software operation is realized and completed before exit.

Other software can also be integrated in a similar way. The comprehensive interfaces of some major professional software implemented are shown in Figures 9.2–9.4.

9.4 Software Development of EM-S-T Field Coupling Analysis

9.4.1 Basic Ideas and Framework

In order to build a scheme of comprehensive software subsystem for field coupling analysis, the top-level design needs to be carried out first, and the specific tasks and requirements, key technologies, and input and output requirements of each part are determined. Secondly, according to the characteristics of field coupling, the analysis process to solve the field coupling problem is determined, and

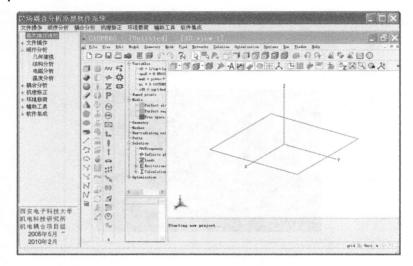

Figure 9.3 Comprehensive integration of electromagnetic analysis software.

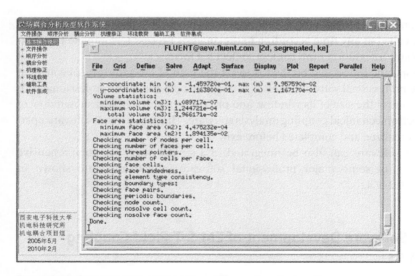

Figure 9.4 Comprehensive integration of thermal analysis software.

a comprehensive human–computer interaction interface is built. Furthermore, on the basis of research on field coupling modeling and solving methods, a data exchange interface with related software or modules has been developed. Finally, the self-developed software modules and existing professional software tools are integrated to form a comprehensive software subsystem [1, 2].

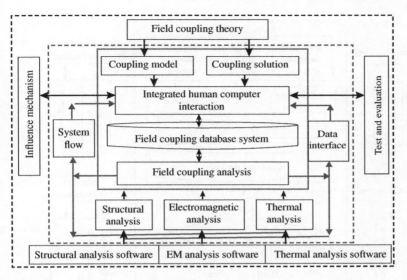

Figure 9.5 The overall framework of the field coupling prototype software.

In the research and development process of this comprehensive software subsystem, a preliminary demand analysis and system design is carried out first. Then, the comprehensive software for typical functions is developed and built. Furthermore, according to the newly established electromechanical coupling model, its functions are expanded and perfected. The existing advanced software tools are integrated, and finally, the software system is formed for numerical analysis of field coupling characteristics. The overall framework of the comprehensive software subsystem is shown in Figure 9.5. Its core is the field coupling model and solution method. Based on the field coupling analysis process, mature commercial software in structural, electromagnetic, and temperature analysis are called when they are needed in the field coupling analysis to improve computational efficiency. The computer configuration of the software system is shown in Figure 9.6, which mainly includes four aspects: system planning, problem estimation, resource allocation, and task scheduling. The relationship between the modules in the prototype software system is shown in Figure 9.7. First, the parameterized module will generate the three-dimensional model of the electronic equipment. After data conversion and model modification, it can be used by the structural analysis module and the electromagnetic analysis module. Meanwhile, conversion can be avoided, and the data can be directly related into the structural analysis module. The traditional analysis method is that the electromagnetic analysis module directly calculates the electromagnetic performance after obtaining the electronic equipment structure, and the software system builds the structural model imported by the

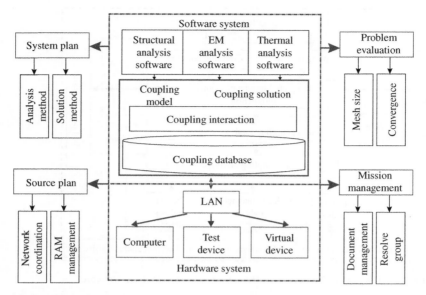

Figure 9.6 Computer configuration structure of field coupling prototype software.

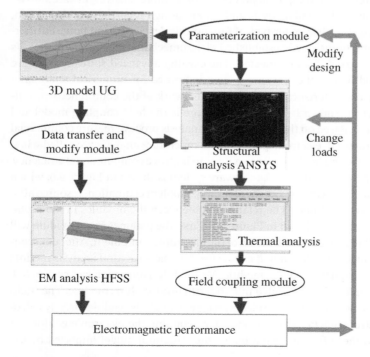

Figure 9.7 Interrelationship of software modules in field coupling prototype software.

parameterized module in the structural analysis module and then performs the temperature field analysis, obtains the temperature distribution of the electronic equipment, and then enters the field coupling module to obtain the electromagnetic performance of the electronic equipment after electromechanical coupling analysis by the field coupling theory and solution method.

9.4.2 Interactive Interface for Field Coupling Analysis

The field coupling analysis process is specifically operated and realized through the interactive interface of the comprehensive software. Therefore, it is necessary to rationally design its operation mode, display mode, calling mode, media mode, etc. and integrate the mechanical structure analysis interface, electromagnetic analysis interface, and thermal analysis interface into the unified field coupling analysis interface for different disciplines. In the process of operation, the operator has unified objects, clear levels, familiar interface forms, and convenient usage. The basic interactive interface structure is shown in Figure 9.8.

9.4.3 Data Exchange Interface

On the basis of the abovementioned work, the problem of data exchanging interface must be solved. In the analysis of structure, electromagnetics, and heat, although the models are different, they have common general database. In this way, the union, intersection, and difference of these different models can be obtained separately. The key indicators and their main influencing factors in

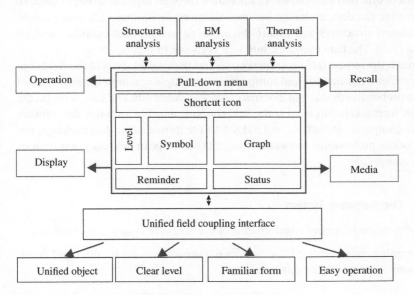

Figure 9.8 Composition of field coupling interaction.

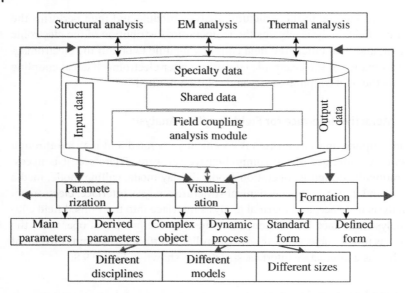

Figure 9.9 Field coupled data exchange interface.

the intersection are determined and parameterized as shared data. The derived parameters in the difference set as professional data are determined. On this basis, the corresponding field coupling analysis database is established. Combining the analysis process of field coupling, a data exchange interface between software modules of different disciplines is established through input and output data, so as to realize the data exchange between different disciplines, different models, and different structures and ensure the achieved smooth field coupling analysis process [3–7]. The data exchange interface is shown in Figure 9.9.

Through the structural, electromagnetic, and thermal analysis of the field coupling problem; the analysis and comparison of multiple common file formats; and the comprehensive evaluation in terms of file size, versatility, and ease of usage, the GEO file format is finally used as the field coupling data file format of the software human–computer interaction, and IGES, SAT file format as the data exchange format between professional software. Thus, data exchange in field coupling analysis can be realized.

9.4.4 The Software System

The software environment requirements of this software system are as follows:

Programming environment: C++ Builder 6.0 version is used for the rapid development of the prototype software.

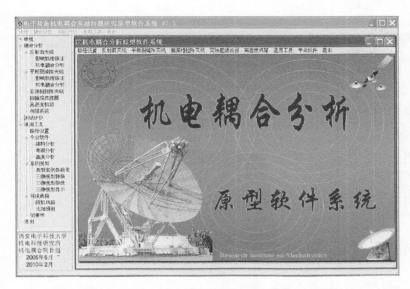

Figure 9.10 The main interface of the software system for field coupling analysis.

Auxiliary programming environment: using MATLAB® R2006b, making full use of its existing function library.

Structural and thermal analysis: using ANSYS Multiphysics V11, taking advantage of its advanced, powerful functions and wide popularity. At the same time, it can support conventional thermal analysis.

Electromagnetic analysis: using FEKO V5.4, using its open data format to realize the import of deformed models.

The above agreement ensures the smooth and rapid integration of the later software modules. The final realization of the main interface of the field coupling analysis software is shown in Figure 9.10, which includes the field coupling module, related professional software, and the comprehensive integration of model modification and conversion modules.

9.4.4.1 Parametric Modeling and Data Processing Module

Parametric modeling and data processing module includes typical case parametric modeling modules, 3D model conversion modules, and 3D model modification modules to achieve heterogeneous collaboration of software tools and improve the usability of software tools. The basic interface forms of these three software modules are shown in Figures 9.11–9.13, respectively.

Figure 9.11 Typical case study module.

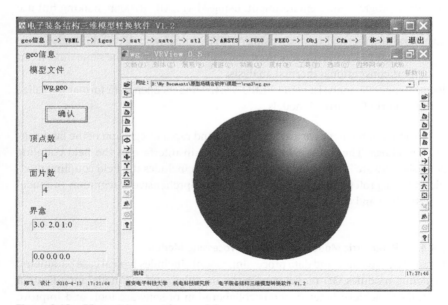

Figure 9.12 3D model conversion module.

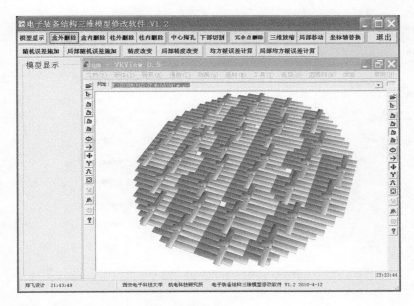

Figure 9.13 3D model modification module.

9.4.4.2 Field Coupling Analysis Module for Several Typical Electronic Equipment

Based on the above work, a number of field coupling analysis modules for typical equipment have been completed. The field coupling analysis modules for reflector antennas, planar slotted array antennas, and active phased array antennas are given below.

Reflector antenna: The reflector antenna field coupling analysis module includes the reflector antenna field coupling analysis module, the antenna wind load automatic generation module, and the antenna solar heat flux generation module (Figures 9.14–9.16).

Planar slotted array antennas: The field coupling analysis module of planar slotted array antennas is shown in Figure 9.17.

Active phased array antennas: The active phased array antenna field coupling analysis module is shown in Figure 9.18.

9.5 Software Development of IM of Structural Factors on Performance

9.5.1 Basic Ideas and Framework

As mentioned before, the effect of mechanical and structural factors on the performance of electronic equipment is relatively complicated. In order to systematically

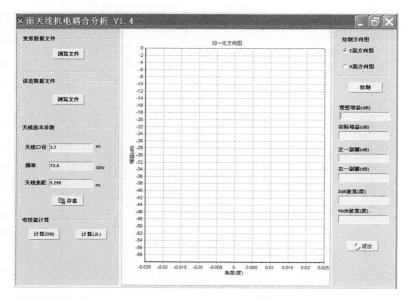

Figure 9.14 Reflector antenna field coupling analysis module.

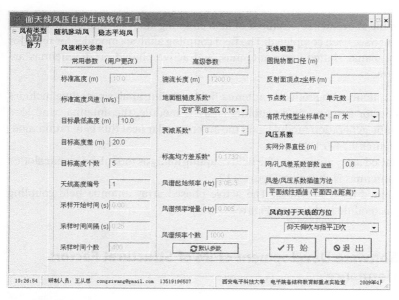

Figure 9.15 Reflector antenna wind load automatic generation module.

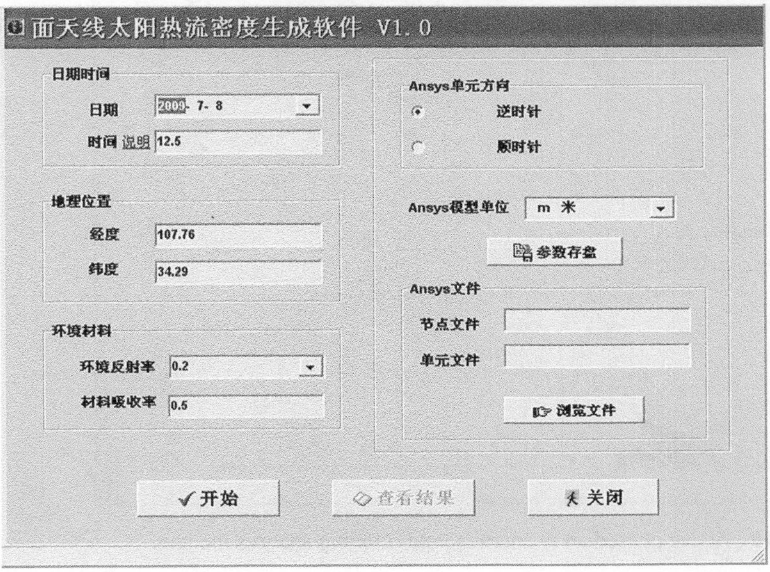

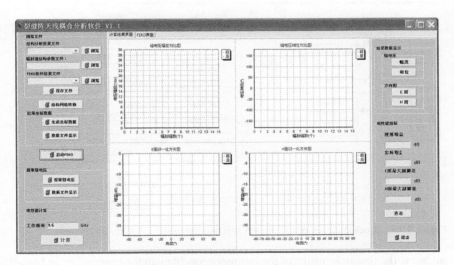

Figure 9.16 Reflector antenna heat flow density generation module.

Figure 9.17 Planar slotted array antenna field coupling analysis module.

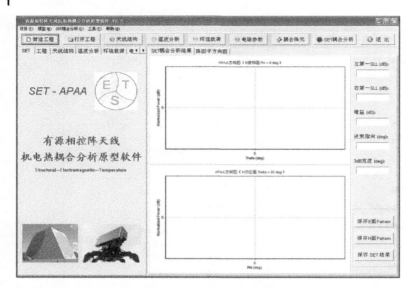

Figure 9.18 Active phased array antenna field coupling analysis module.

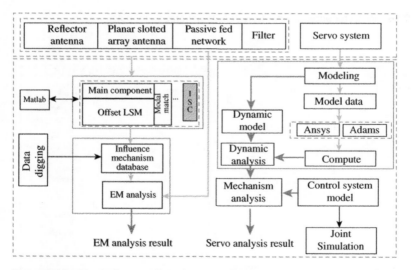

Figure 9.19 Block diagram of the impact mechanism prototype software system.

integrate the research results of the influence mechanism for application and promotion, the software system adopts the form shown in Figure 9.19, with functions such as influence mechanism analysis, empirical formula fitting, and electrical performance analysis.

9.5.2 The Software of the Influence Mechanism of the Antenna Feeder System

Different from the traditional analysis method, in the analysis process of the influence mechanism of the antenna feeder system, the influence mechanism analysis model based on the intermediate electric parameter is adopted, that is the influence relationship of the structural factors on the intermediate electric parameter is stored in the database in the form of data model [8–16]. When analyzing the electrical properties, the corresponding intermediate electrical parameter values are firstly obtained according to the structural parameters and manufacturing accuracy, and secondly, the traditional method is used to calculate the electrical performance according to the intermediate electrical parameter values.

The antenna feeder system analysis interface is shown in Figure 9.20, which is mainly composed of two modules. The first module is the analysis module of impact mechanism, which can carry out principal component analysis and partial least squares analysis on the original data in the impact mechanism data warehouse and feedback the analysis results to the data mining system for data mining, so as to form the influence mechanism database. The relational model (data model) of the influence of structural factors on the intermediate electrical performance is stored in this database, which can be predicted by support vector machine, that is the reasonable estimation of the corresponding intermediate electrical parameters (such as the admittance array of planar slotted array antennas,

Figure 9.20 Analysis interface of antenna feeder system.

the unloaded Q value of the filter, and coupling coefficient) is given according to the given structural factors. The second module is the electrical performance analysis module. The electrical performance is obtained using the equivalent circuit method according to the admittance array of the planar slotted array antennas, the unloaded Q value, and the coupling coefficient of the filter, so as to analyze the influence of typical structural factors in reflector antennas, planar slotted array antennas, and filter on the electrical performance.

9.5.3 The Software of Servo System's Influence Mechanism

The analysis of the influence mechanism of the servo system is based on the system dynamics equation of comprehensive structural factors, and the parameterized modeling and automatic parameter extraction methods are used to obtain the model parameters corresponding to the dynamic equations, and finally, the dynamics analysis and servo tracking performance analysis are performed [17–24]. Correspondingly, the software system includes four parts: servo system topology management, structural parameter calculation, dynamic characteristic analysis, and servo tracking performance analysis.

9.5.3.1 Topological Structure Management of Radar Antenna Servo System

Servo system topology management can realize the establishment and maintenance of the servo system topology library. Through this module, the parametric modeling of the gear transmission chain can be realized. Analysis software such as structural parameter calculation modules ADAMS and ANSYS can provide the inertia, stiffness, and other properties of the structure. The controller selection and control parameter setting module are used to specify the controller type of the speed loop and position loop and set the corresponding control parameters. When parameter extraction and control parameter setting are completed, simulation analysis can be carried out, including dynamic characteristic analysis and unit step response analysis. The result of the dynamic characteristic analysis is the torsional vibration mode of each order of the transmission chain, and the result of the unit step response analysis is the position response, tracking error and servo performance estimation.

The details of the servo topology database are depicted in Figure 9.21, where the topology defined in the database is given in the upper left table. The upper right table shows the mechanism sketch for the selected topology. The lower two tables provide the gear information and the drive shaft information for this topology, respectively. This module allows the user to view the details of all topologies in the database. In this interface, users can do the following operations: (i) Add new structure: this function can realize the definition of topology. Users can add new topology and save its mechanism sketch to the database. (ii) Add Gear: select

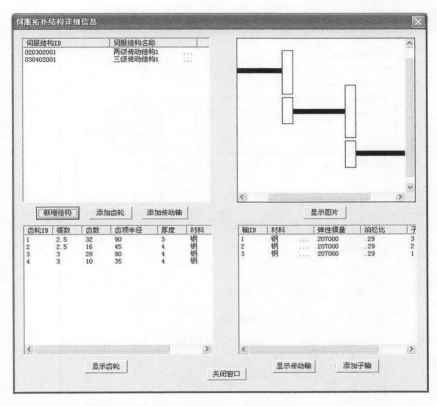

Figure 9.21 Radar antenna servo topology management.

structure in the topology list, click "Add Gear," in the pop-up dialog box, users can define gear information, and the gear identification number is automatically generated by the system. (iii) Add Drive Shaft: select structure in the topology in the pop-up dialog box, the user can define the basic information of the drive shaft, and the identification number of the drive shaft is automatically generated by the system. (iv) Add subshaft: considering that the drive shaft may be a step shaft, the parameters of each section of the drive shaft need to be defined in detail. Figure 9.22 shows the process of defining the subshaft, and the user can define the subshaft keyway and other dimensional parameters.

9.5.3.2 Calculation of Structural Parameters

Structural parameters are calculated by invoking analysis software such as ADAMS and ANSYS to calculate the dynamics of the servo structure and prepare for the calculation of the servo system influence mechanism. After setting up

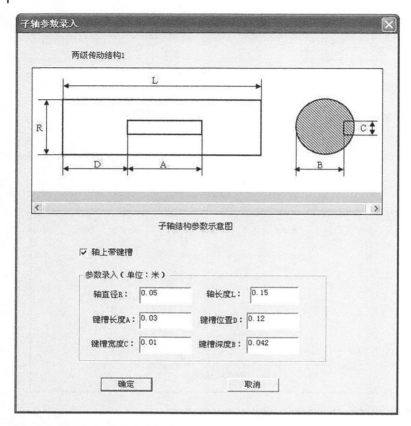

Figure 9.22 Subaxis definition process.

the ADAMS and ANSYS installation path, selecting the template to be applied, and setting the simulation work path, the geometric and mechanical parameters required for the influence mechanism simulation can be automatically calculated and extracted. Figure 9.23 shows the parameter extraction interface.

9.5.3.3 Controller Selection and Control Parameter Setting
The software sets up an interface for selecting the structure form of the position and velocity loop controllers and setting the corresponding control parameters. The user-selected controller form and the corresponding control parameters are automatically stored in the database for recall during the influence mechanism analysis. The corresponding software interface is shown in Figure 9.24.

9.5.3.4 Influence Mechanism Analysis
When the parameter extraction and control parameter setting are completed, simulation analysis can be carried out, including dynamic characteristic analysis and

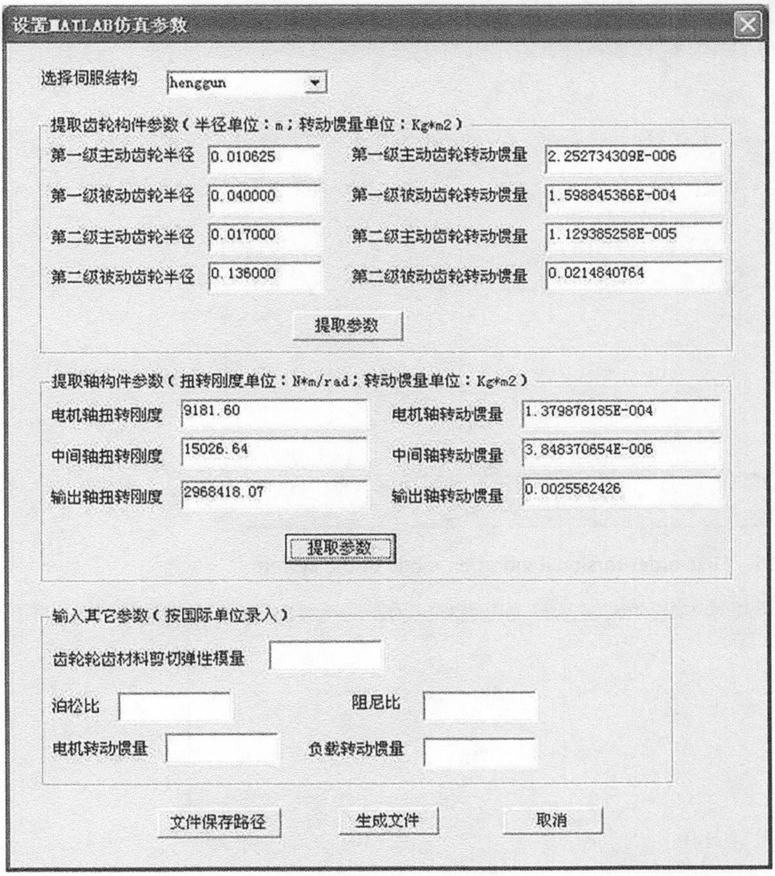

设置MATLAB仿真参数
选择伺服结构
henggun
提取齿轮构件参数（半径单位：m；转动惯量单位：Kg*m2）
第一级主动齿轮半径
0.010625
第一级主动齿轮转动惯量
2.252734309E-006
第一级被动齿轮半径
0.040000
第一级被动齿轮转动惯量
1.598845366E-004
第二级主动齿轮半径
0.017000
第二级主动齿轮转动惯量
1.129385258E-005
第二级被动齿轮半径
0.136000
第二级被动齿轮转动惯量
0.0214840764
提取参数
提取轴构件参数（扭转刚度单位：N*m/rad；转动惯量单位：Kg*m2）
电机轴扭转刚度
9181.60
电机轴转动惯量
1.379878185E-004
中间轴扭转刚度
15026.64
中间轴转动惯量
3.848370654E-006
输出轴扭转刚度
2968418.07
输出轴转动惯量
0.0025562426
提取参数
输入其它参数（按国际单位录入）
齿轮轮齿材料剪切弹性模量
泊松比
阻尼比
电机转动惯量
负载转动惯量
文件保存路径
生成文件
取消

Figure 9.23 Simulation calculation and extraction.

设置仿真对象
选择伺服结构
henggun
选择控制器类型
参数文件保存路径
确定
取消
设置仿真参数
Ref_Pos
Actual_Pos
Pos_out
位置环控制器
Ref_Velocity
Actual_Velocity
Velocity_out
速度环控制器
Ref_Velocity
Position
Velocity
伺服机构
位置环参数
速度环参数
伺服机构参数
生成参数文件
关闭

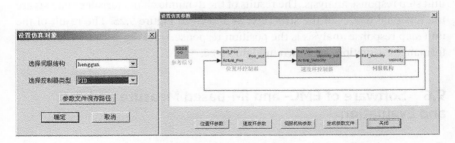

Figure 9.24 Controller selection and control parameter setting interface.

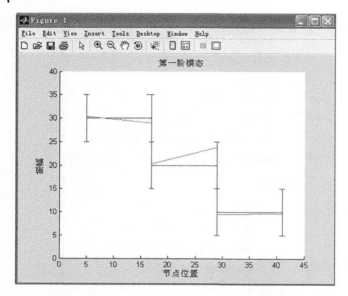

Figure 9.25 First-order torsional vibration mode of the system.

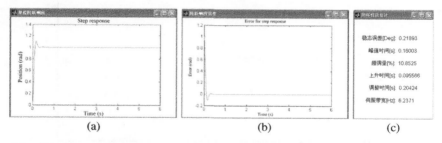

Figure 9.26 Unit step response of the system. (a) Step response, (b) Error for step response, (c) performance.

unit step response analysis. The results of the dynamic characteristics analysis are the torsional modes of the structure, as shown in Figure 9.25. The result of the unit step response analysis is the position response, tracking error and servo performance estimation, as shown in Figure 9.26.

9.6 Software of EMC- and IM-based Measure and Test and Evaluation

The comprehensive testing and evaluation of electromechanical coupling is a comprehensive embodiment of the results of electromechanical coupling theory and influence mechanism of electronic equipment. Based on the concept of electromechanical coupling, the comprehensive test and evaluation prototype system, which is hoped to be constructed, has the following characteristics. First, advance, that is using advanced technology (computer network technology, database technology, etc.) to build an advanced test platform. Second, scalability, that is in order to cover the next generation of electronic equipment testing and evaluation needs, the system must have good scalability. The third is practicality, that is the main purpose is to test the coupling theory and influence mechanism, and the technology used and the functions realized are closely focused on this purpose.

9.6.1 Basic Idea and Framework

The general framework of the comprehensive test and evaluation system is shown in Figure 9.27, which includes "module of coupling analysis among the test factors," "typical case test platform interface module," "comprehensive evaluation module," and "field coupling analysis with access module of influence mechanism," in addition to a public SQL Server database platform.

The module of coupling analysis among the test factors is used to analyze what among the structural parameters affecting electrical performance are the major ones (must be tested), and what are the minor ones (can be considered not to be

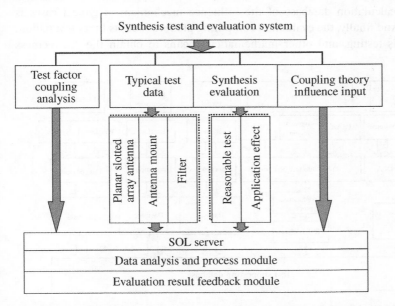

Figure 9.27 Components of the comprehensive assessment prototype system.

tested). The typical case test data interface module realizes the data interface with the specific case test platform and completes the input and collection of structural and electrical performance parameters for each case [25, 26].

The field coupling analysis and influence mechanism access module is responsible for the data interface between these two prototype systems and provides theoretical calculation data for comprehensive evaluation.

Evaluation result feedback module is developed to complete the correctness and effectiveness test of the field coupling analysis, influence mechanism.

9.6.2 The Working Process

Figure 9.28 shows the workflow of the comprehensive measurement system. Firstly, through the coupling degree analysis based on subjective and objective coupling, the coupling degree database is formed to provide the basement of determining the structural parameters to be tested. Secondly, the test data of typical cases are obtained through the public interface of typical cases, and the typical case test database is formed after necessary processing. Thirdly, through the coupling theory interface to obtain the calculation data of the coupling theory on typical cases, a database of the calculation of the coupling theory of typical cases is formed. Fourthly, through the influence mechanism interface, the calculation data of the influence mechanism on typical cases are obtained, and the calculation database of the influence mechanism of typical cases is formed. And finally, the evaluation computing platform adopts fuzzy evaluation, hypothesis testing, and other mathematical means to obtain the "correctness

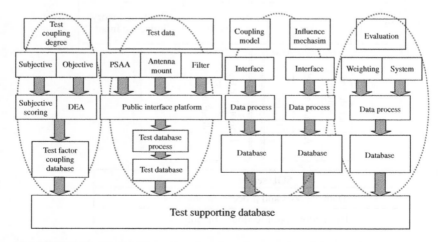

Figure 9.28 Workflow of the comprehensive assessment prototype system.

test" and "effectiveness evaluation" of the "coupling theory" and "influence mechanism." The conclusion of "effectiveness evaluation" can be obtained.

9.6.3 Database

The SQL Server database used here is a symmetrical multiprocessor structure, with preemptive multitasking management, perfect security system and fault tolerance and it is easy to be maintained. The SQL Server uses two levels of security authentication, login authentication and permission verification of database user accounts and roles to ensure the security of the database.

Figure 9.29 shows the main framework of the database of this prototype system. The main databases include the typical case structure parameter database, the typical case electrical performance parameter database, the typical case coupling degree database, the typical case evaluation system database, the typical case electrical performance parameter (theoretical calculation) database, and other databases. Each database consists of a number of specific data tables, which are used to form a relational database by primary and foreign keys.

9.6.4 Test Data Interface

A large amount of test data needs to be entered in this comprehensive evaluation system, and for this reason, the settings and operations of the test data interface

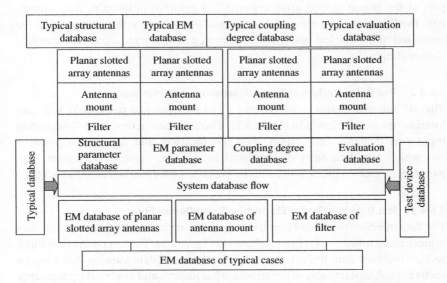

Figure 9.29 Database architecture of the comprehensive assessment prototype system.

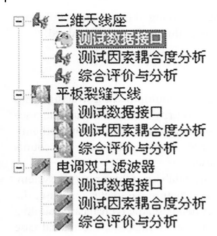

Figure 9.30   Test data interface tree diagram.

are described below [27–29]. Selecting "Test Data Interface" in the tree on the left side of the main interface of the software (as shown in Figure 9.30) for entering the module.

9.6.4.1   Test Data Interface for Planar Slotted Array Antennas

The data interface module of the structural and electrical performance parameters of the planar slotted array antenna is shown in Figure 9.31. The data input part of the planar slotted array antenna test parameters includes data import, data modification, data view, and other modules to realize the data entry of the structural and electrical performance parameters of the radiation slot/coupling slot/excitation slot.

9.6.4.2   Test Data Interface of Three-dimensional Antenna Base

The 3D antenna mount structure and servo performance parameters test data interface module is shown in Figure 9.32. The 3D antenna mount test parameters input part includes data import, data modification, and data viewing, which can realize the data entry of structural parameters, servo performance, gear parameters, inertia/stiffness of azimuth axis/pitch axis/transverse roll axis, etc.

9.6.4.3   Test Data Interface of Electrically Tuned Duplex Filter

The data interface module of the structure and electrical performance parameters of electrically tuned duplex filter is shown in Figure 9.33. The test parameters input section includes data import, data modification, and data viewing. Clicking on each tab enables data entry of the structure parameters and electrical performance parameters such as coupling holes/coupling rings/inner conductors.

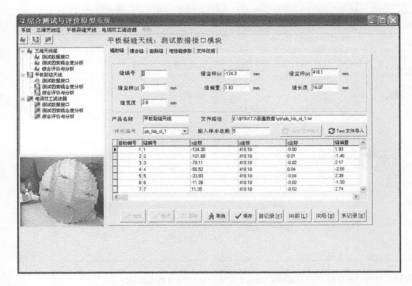

Figure 9.31 Test data interface module of planar slotted array antennas.

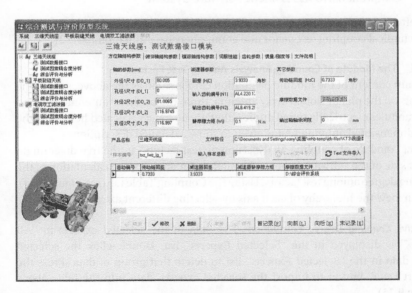

Figure 9.32 Three-dimensional antenna base test data interface module.

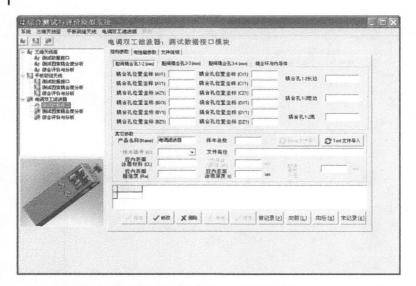

Figure 9.33 Interface of electrically tuned filter structure parameters.

9.6.5 Comprehensive Assessment Software System

The software system consists of two main modules, i.e. coupling analysis of electromechanical coupling test factor and comprehensive evaluation [30–32].

9.6.5.1 The Module of Coupling Analysis Among Test Factors

The main content of the module pagination card is same as the above, so that the contents on the planar slotted array antenna, three-dimensional antenna seat, and three modules electrically tuned duplex filter are no longer repeated here; only the three-dimensional antenna seat is given to show a detailed description.

Select "module of coupling analysis among test factors" in the tree diagram on the left side of the software to enter the module for analysis and calculation of the coupling degree among test factors. Using the Coupling Table tab (see Figure 9.34), one can view the hierarchy of each axis by selecting different axes.

Subjective Coupling Degree Calculation The data of the selected experts in "All Experts" is displayed in the "Selected Experts" list. Double-click the selected expert data in the "Selected Experts" list to delete that group of data. Press the "Import Data" button to import the selected experts' data into the table above (Figure 9.35).

After importing the data, press the "Coupling Calculation" button to calculate the selected data and display the calculation results in the table above the button (Figure 9.36).

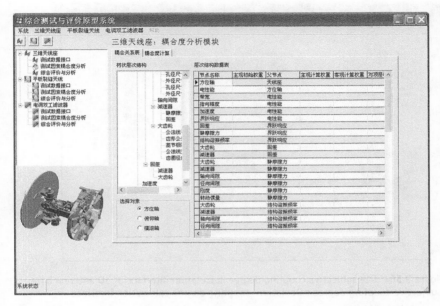

Figure 9.34 Three-dimensional antenna base coupling relationship table interface.

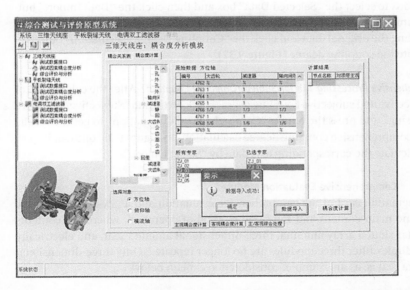

Figure 9.35 3D antenna base subjective coupling degree calculation raw data import interface.

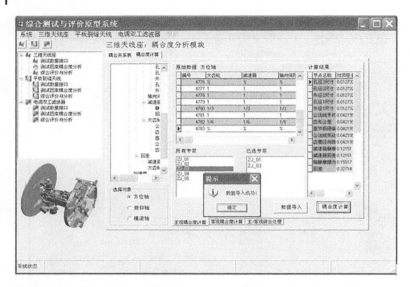

Figure 9.36 The subjective coupling degree calculation interface of 3D antenna base "subjective coupling degree calculation" coupling degree calculation interface.

Objective Coupling Degree Calculation Double-click the data file name in the "All Data" box to select the "Selected Data" box and then click the "Data Import" button to import the original data. After importing the original data, click the "After" importing the original data and click the "Coupling Degree Calculation" button to calculate the coupling degree (Figure 9.37).

Comprehensive Processing Tab for Subject/Objective Coupling After the calculation of the objective and subjective coupling degree, click on the tab "Subjective/Objective Processing" and press the button "Comprehensive Calculation" to get the results of the comprehensive coupling degree calculation. In Figure 9.38, one can import and calculate the corresponding axis data by "Select Object."

9.6.5.2 Comprehensive Evaluation Module
The pagination card of the comprehensive evaluation and analysis module have the same main content and same method of operation, where the contents in planar slotted array antenna, three-dimensional antenna seat, and electrically tuned duplex filter three modules are no longer repeated. Only three-dimensional antenna seat is, as an example, considered to explain below.

Correctness Check Tab To perform a correctness check, click on the "Correctness Check" tab of the corresponding "Comprehensive Assessment Analysis" module

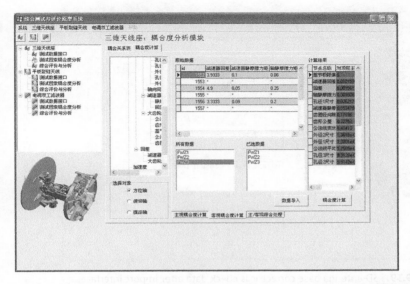

Figure 9.37 Three-dimensional antenna seat objective coupling degree calculation interface.

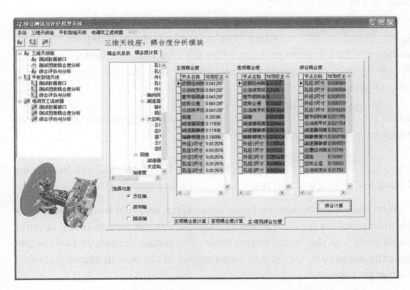

Figure 9.38 3D antenna base subjective/objective comprehensive processing comprehensive calculation interface.

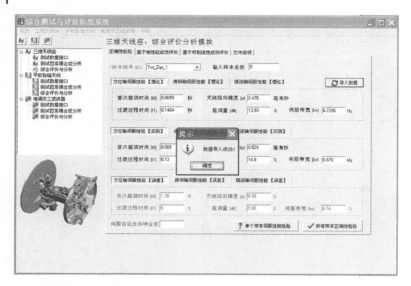

Figure 9.39 3D antenna base correctness check data after import interface.

to access it. Enter the total number of samples and click on the "Sample number" drop-down box to select the sample.

(1) Click _____ 导入数据 _____. Click "Yes" to confirm the data import, that is to complete the data import operation in the selected sample number, the system will pop up the "data import success" prompt box after the successful data import (Figure 9.39).

(2) Click _____ ? 单个样本伺服性能检验 _____, then the overall agreement of the servo performance of this sample is obtained.

(3) Click _____ ✓ 所有样本正确性检验 _____, the "Servo Correctness Check Results for All Samples" dialog box will pop up, as shown in Figure 9.40.

This screen shows the maximum error, mean error, and root mean square error of the parameter values for each axis in all sample data, as well as the maximum error, mean error, and root mean square error of all parameter values, and the last line shows the maximum, mean, and mean square of the overall agreement of all sample servo performance.

Based on the Electrical Performance Evaluation Tab   Click the "Electrical Performance Based Evaluation" tab of the "Comprehensive Assessment and Analysis" module to enter (Figure 9.41). Enter the total number of samples and click on the "Sample

Figure 9.40 3D antenna base all samples servo correctness check results interface.

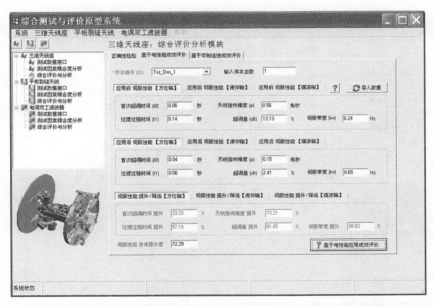

Figure 9.41 Sample interface of 3D antenna mount based on electrical performance effectiveness test.

Number" drop-down box to select samples. (When the input sample number is larger than the actual number of samples, a warning box will pop up, click "Yes," then the total number of input samples will be set to empty; click "No," then the total number of input samples will be set to the actual number of samples.)

(1) The import data button operates as before.

(2) Click ⟨**? 基于电性能应用成效评价**⟩, the overall servo performance improvement degree for this sample is obtained.

Based on the Manufacturability Effectiveness Evaluation Tab Click on the "Evaluating the Effectiveness of Manufacturability" tab in the "Comprehensive Evaluation Analysis" section (Figure 9.42). Enter the total number of samples and click on the "Sample Number" drop-down box to select a sample.

Import New Data

(1) Click ⟨**🗘 Txt文件导入**⟩, confirm the import in the pop-up dialog box to complete the data importing (Figure 9.43).

(2) Clicking the "Modify" button can make changes to the data (Figure 9.44)

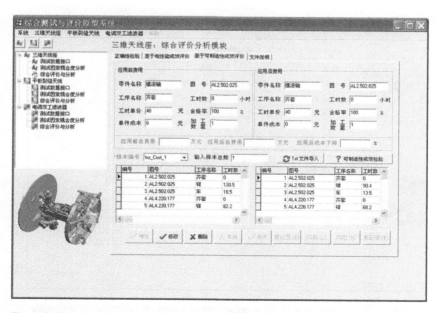

Figure 9.42 Sample interface for evaluating the manufacturability-based effectiveness of three-dimensional antenna mounts.

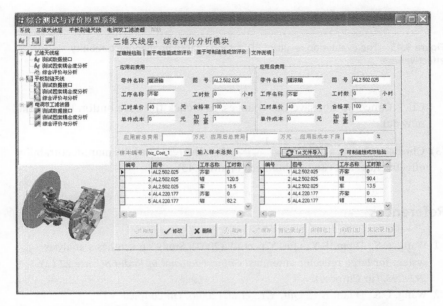

Figure 9.43 Interface after importing sample data of 3D antenna holders based on manufacturability effectiveness evaluation.

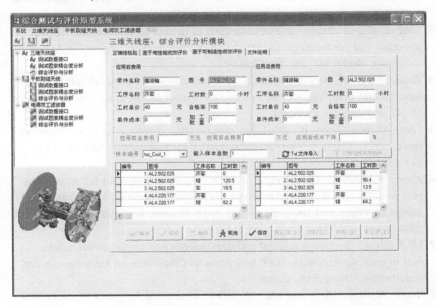

Figure 9.44 Sample data modification interface for 3D antenna holders based on manufacturability effectiveness evaluation.

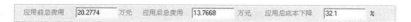

Figure 9.45 The evaluation result of 3D antenna based on manufacturability effectiveness.

Click Cancel to not save the changes, and click the Save button to save the new data.

(3) Click ?可制造性成效检验 , then the results of the manufacturability effectiveness test can be obtained (Figure 9.45).

References

1 Wang, C.S., Duan, B.Y., Qiu, Y.Y. et al. (2007). Study on synthesis analysis system for large reflector antennas. *Chinese Journal of Radio Science* 22 (2): 292–298. (in Chinese).

2 Wang, C.S., Duan, B.Y., Qiu, Y.Y., et al. (2008). On coupled structural-electromagnetic optimization and analysis of large reflector antennas. *The 8th International Conference on Frontiers of Design and Manufacturing (ICFDM2008)*, September 23–26, 2008, Tianjin, China.

3 Song, L.W. and Zheng, F. (2009). Analysis of the coupled problems between structure and electromagnetism based on discrete meshes. *Journal of Xidian University* 36 (2): 347–352. (in Chinese).

4 Zheng, F. (2006). Barycentric coordinate mapping and its application in uniform texture mapping. *Journal of Computer-Aided Design & Computer Graphics* 18 (4): 482–486. (in Chinese).

5 Chen, M. and Zheng, F. (2006). Improved texture mapping based on improved mesh parametrization. *Journal of Computational Information Systems.* 2 (2): 645–650.

6 Zheng, F. (2009). Effective uniform mesh parameterization for boundary complex models. *International Conference of Computer Graphics, Visualization, Computer Vision and Image processing 2009*, Algarve, Portugal, June 20–22, 2009

7 Li, P. (2008). Electromechanical coupled analysis and experimental validation of reflector antennas. *IEEE International Conference on Mechatronics and Automation*, Changchun, China.

8 Wang, W., Duan, B.Y., and Ma, B.Y. (2008). A method for panel adjustment of large reflector antenna surface and its application. *Acta Electronica Sinica* 36 (6): 1114–1118. (in Chinese).

9 Wang, W., Li, P., and Song, L.W. (2009). Mechanism of the influence of the panel positional error on the power pattern of large reflector antennas. *Journal of Xidian University* 36 (4): 708–713. (in Chinese).

10 Wang, W., Wang, C.S., and Li, P., et al. (2008). Panel adjustment error of large reflector antennas considering electromechanical coupling. *IEEE/ASME International Conference on Advanced Intelligent Mechatronics (AIM2008)*, August 2–5, 2008, Xi'an, China.

11 Zhou, J.Z., Duan, B.Y., and Huang, J. (2009). Predication of plane slotted-array antenna electrical performance affected by manufacturing precision. *Journal of University of Electronic Science and Technology of China* 38 (6): 1047–1051. (in Chinese).

12 Yu, W., Gu, W.J., and Guo, X.S. (2008). Pattern analysis for deformed subarray of planar slotted antenna array. *Modern Radar* 30 (12): 70–73. (in Chinese).

13 Yan, Z.J. (2009). Effect of mechanics structure on the electric performances of radiation element in planar slotted antenna. *Telecommunication Engineering* 49 (6): 60–65. (in Chinese).

14 Xiong C W and Wang Y. Analysis and computation of microwave devices surface RF equivalent conductivity. *Academic Conference on Electromechanical and Microwave Structure Technology*, 2008.8.1–5, Jiang Xi, Jiu Jiang. (in Chinese)

15 Zhou, J.Z. and Huang, J. (2010). Incorporating priori knowledge into linear programming support vector regression. *2010 IEEE International Conference on Intelligent Computing and comprehensive Systems (IEEE ICISS2010)*. October 22–24, 2010, Guilin, Guangxi, China.

16 Zhou, J.Z., Duan, B.Y., and Huang, J. (2010). Influence and tuning of tunable screws for microwave filters using least squares support vector regression. *International Journal of RF and Microwave Computer Aided Engineering.* 20 (4): 422–428.

17 Li, S.L., Huang, J., and Duan, B.Y. (2010). Comprehensive design of structure and control for radar antenna servo-mechanism. *Journal of Mechanical Engineering* 46 (1): 140–146. (in Chinese).

18 Bao, H., Duan, B.Y., Du, J.L. et al. (2008). Synchronous optimization design of control and structure of complex mechanism. *Chinese Journal of Computational Mechanics* 25 (1): 8–13. (in Chinese).

19 Tong, X.F., Huang, J., and Zhang, D.X. (2009). Multidisciplinary joint simulation technology for servo mechanism analysis. *2009 IEEE International Conference on Information and Automation*, 6: 655–658.

20 Zhou, J.Z., Duan, B.Y., and Huang, J. (2008). Effect and compensation for servo systems using LuGre friction model. *Control Theory & Applications.* 25 (6): 990–994. (in Chinese).

21 Zhou, J.Z. and Huang, J. (2009). Support vector regression modeling and back-stepping control of friction in servo system. *Control Theory & Applications* 26 (12): 1405–1408. (in Chinese).

22 Huang, J. (2008). Backlash compensation in servo systems based on adaptive backstepping-control. *Control Theory & Applications* 25 (6): 1090–1094. (in Chinese).

23 Zhou, J.Z. and Huang, J. (2009). Modeling of servo system with backlash and its influence on open-loop frequency characteristics. *China Mechanical Engineering* 20 (14): 1721–1725. (in Chinese).

24 Huang J. (2009). HLA based multidisciplinary joint simulation technology for servo mechanism analysis. *2009 IEEE International Conference on Mechatronics and Automation*, 8: 4517–4522.

25 Ma, H.B. and Chen, G.D. (2008). The method for computing degree of testing factors in electromechanical coupling research. *Proceedings of the Annual Conference of the Electronic Mechanical Engineering Branch of the Chinese Institute of Electronics*, 11, Tai Yuan of China (in Chinese)

26 Shen, Z.F. (2006). Electromechanical coupling analysis of testing factors of electrically tuned duplex filter. *The first academic conference of the research project on the basic problems of electromechanical coupling of electronic equipment*, 11.10–11, Cheng Du of China. (in Chinese)

27 Jia, C.G., Tao, Z.J., Zhang, S. et al. (2009). On dual cameras synchronization for high speed photogrammetry system. *Journal of Geomatics Science and Technology* 25 (6): 76–78. (in Chinese).

28 Shen, Z.F. (2009). Detection simulation and optimization design of electrically tuned duplex filter. *Defense Technology* 5: 52–55. (in Chinese).

29 Zhou, J.Z., Zhang, F.S., Huang, J. et al. (2010). Computer-aided tuning of cavity filters using kernel machine learning. *Acta Electronica Sinica* 38 (6): 1274–1278. (in Chinese).

30 Ma, H.B., Yang, D.W., and Zhou J Z. (2010). Improved coupling matrix extracting method for Chebyshev coaxial-cavity filter. *Progress in Electromagnetics Research Symposium*, March 22–26, 2010, Xi'an, China.

31 Gu, J.F., Ping, L.H., and Liu, J.C.. Coupled vibration analysis for test sample resonating with vibration shaker. *The Third Asia International Symposium on Mechatronics (AISM2008)*, August 27–31, Sapporo, Japan.

32 Zhen, L.D. (2009). A study on fabrication of electrically tuned duplex. *Electro-Mechanical Engineering* 25 (5): 46–48. (in Chinese).

10

Engineering Applications of EMC Theory and IM of Electronic Equipment

10.1 Introduction

In the abovementioned parts of this book, the theoretical model of electromechanical–thermal field coupling for electromechanical coupling analysis of electronic equipment, the influence mechanism of structural factors (including structural parameters and manufacturing accuracy) on electrical performance, and the strategies and methods of electromechanical coupling design based on field coupling theory and influence mechanism are proposed, and finally, the above results are solidified in a comprehensive design platform integrating electromagnetic, structural, thermal, and control for engineering applications. Several typical engineering applications of electronic equipment are presented below.

10.2 Application of Moon-exploration Antenna with the Diameter of 40 m

This antenna is one of the ground systems for lunar exploration project, installed on the Phoenix Mountain in Kunming city of China, and has successfully completed the task of receiving information and measuring orbit of Chang'e 1 and 2 satellites. The antenna is required to have a reflector weight of no more than 70 tons, an efficiency of not less than 0.6, the first sidelobe level of less than −18 dB, and the surface error of less than 0.8 mm. The antenna backup structure consists of 32 radiating beams and a central body, which as a whole is shown in Figure 10.1, and the other structural and electrical parameters are shown in Section 8.2.

There are two key issues addressed: first, electromechanical coupling analysis and design, and second, reflector surface homogeneous design in the elevation angle range of 5°–95°. The conventional design has two problems, i.e. the requirement on self-weight and electrical efficiency of the reflector are difficult to be

Electromechanical Coupling Theory, Methodology and Applications for High-Performance Microwave Equipment, First Edition. Baoyan Duan and Shuxin Zhang.
© 2023 The Institute of Electrical and Electronics Engineers, Inc. Published 2023 by John Wiley & Sons, Inc.

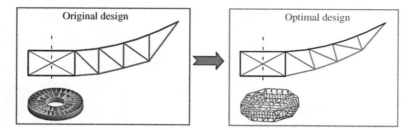

Figure 10.1 Comparison of old and new 40 m reflector antenna solutions.

satisfied simultaneously as one, while as the other, the elevation range of 5°–95° makes it difficult to meet the 0.8 mm reflector precision requirement.

To solve these problems, a new solution is obtained by applying the electromechanical coupling theory, methods, and software. By optimizing the structural design variables such as size, shape, topology, and type, an optimum distribution of the overall stiffness in the electrical performance sense is achieved (Figure 10.1) [1–5]. Not only does it result in a 16.86% reduction in self-weight with improved electrical performance, but also it ensures reflector precision of 0.8 mm during the whole elevation range of 5°–95°.

10.3 Application of the Servomechanism of the Gun-guided Radar System in Warship

Figure 10.2 shows a naval close-range antimissile weapon system, as the last line of defense for a warship, where the failure to intercept would result in the destruction of the entire warship. In addition to requiring its weight to be no more than 700 kg and an antenna mount with an inherent frequency higher than 24 Hz, the radar is required to have an angular velocity of 90 and an angular acceleration of 150, with 1 mrad pointing accuracy, under the violent impact of 4200 rounds per minute. This provides an unprecedented challenge to the design of the antenna mount and servo system.

The key issues addressed by the electromechanical coupling design are, first, the conflict between speed and light weight and, second, effective vibration isolation between the radar and the gun.

The electromechanical separation design has three problems: firstly, it is too heavy, reaching 850 kg, exceeding the 700 kg limit; secondly, the inherent frequency is low, only 22.5 Hz, making the response speed unacceptable; and thirdly, the vibration isolation performance is poor, and the tracking accuracy is low. The tracking accuracy can only be 1.6 mrad under a 15 g shock, while if it

Figure 10.2 Physical diagram of the fire control radar structure of a naval close-range antimissile weapon system. Source: Hoa Tran/VCCorp. Joint Stock Company.

is required to have a tracking accuracy of 1 mrad, it can only withstand a shock acceleration of 9.2 g.

To solve these problems, the theory, method, and software of electromechanical coupling design are applied by optimizing the structural parameters and controlling the gain factor. The optimum distribution of the overall mass, inertia, and stiffness of the structure is achieved with increasing the inherent frequency and simplifying the transmission system and vibration isolation system [6–14]. Compared to the original solution, three benefits are achieved: firstly, the self-weight has been reduced by 41% to 500 kg; secondly, the inherent frequency has been increased by 52% to 34.2 Hz; and thirdly, a tracking accuracy of 0.5 mrad has been achieved at an impact of 15 g (Figure 10.3).

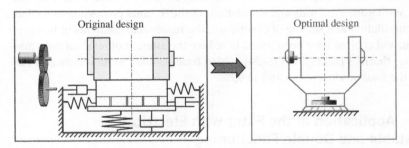

Figure 10.3 Comparison of the traditional and new fire control radar structure of a ship-based close-range antimissile system.

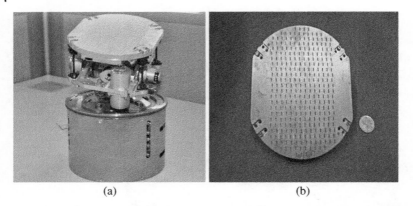

(a)   (b)

Figure 10.4 Schematic diagram of the antenna array structure. (a) side view, (b) top view.

10.4 Application of Planar Slotted Array Antennas

Figure 10.4 shows a millimeter-wave enemy identification system in which one of the important components is the planar slotted array antenna, operating in the Ka frequency band.

The main problem with the traditional design is the low yield rate of 75%. The main factor for the low yield is the complex welding process. Two main issues are addressed: one is that electrical properties are sensitive to manufacturing accuracy such as radiation slots, and the law of influence needs to be found, and the second is how to improve the yield of the welding.

Through the application of electromechanical coupling theory, methods, and the comprehensive software platform, the main parameters and processes affecting the electrical properties have been explored. The reasonable jig application positions have been found to reduce false welding. The methods and means of using laser cutting solder have been applied so that different machining accuracy requirements can be targeted, and the machining difficulty is reduced to a certain extent. With numerical tests and simulations of the thermal processing as well as the accumulation and sorting of experience, the reasonable gradients of heating, holding, and cooling have been found to reduce the amount of residual deformation after thermal processing [15–24], which have resulted in an increase in the rate of the final products from 75% to 95%.

10.5 Application of the Filter with Electrical Adjustable and Double Functioning

The filter with electrical adjustable and double functioning is a key device for scattering telecommunications, operating in the frequency from 610 to 960 MHz,

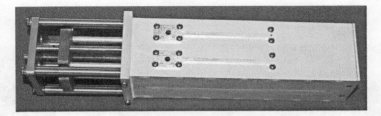

Figure 10.5 Physical model of the filter with electrical adjustable and double functioning.

requiring stable tuning over the full frequency band with constant bandwidth and out-of-band suppression. The structure uses a coaxial cavity-type sequential coupling, and the tuning is achieved by adjusting the length of the inner conductor. As the tunable frequency range is relatively wide, the interstage coupling adopts the way of large-hole coupling, and the input–output coupling uses the trapezoidal coupling ring structure near the inner conductor, whose physical model is shown in Figure 10.5.

The filter has the problem of extremely high requirements for internal wall finish and form tolerance indicators, which are very difficult to meet with actual machining. For this reason, in practical engineering, the traditional approach is to make several more backups and let the ESC decide which one to be choose. It leads to the problem of long period time, high cost, and difficult to guarantee performance in filter development. To solve the problem, the electromechanical coupling theory, methods, and software tools, especially the mechanical mechanism factors on the electrical performance of the mechanism, are applied to find the main impact parameters and processing accuracy [25–32], greatly improving the debugging efficiency, which is from the original four days to a half-day debugging one.

10.6 Application of FAST-500 M Aperture Spherical Radio Telescope

As shown in Figure 1.8, FAST 500 m aperture spherical radio telescope, located in Pingtang of Guizhou Province, Southwest part of China, is the world's largest single radio telescope. There are two main challenges. One is the ultralight weight mechanism (structure) with high frequency band, large bandwidth, and high precision. The other is how to realize a millimeter dynamic positioning accuracy of the feed supported and driven with long and flexible cables. The repeated design and tests have shown that the traditional electromechanical separation design is powerless. By applying electromechanical coupling theory, method, and

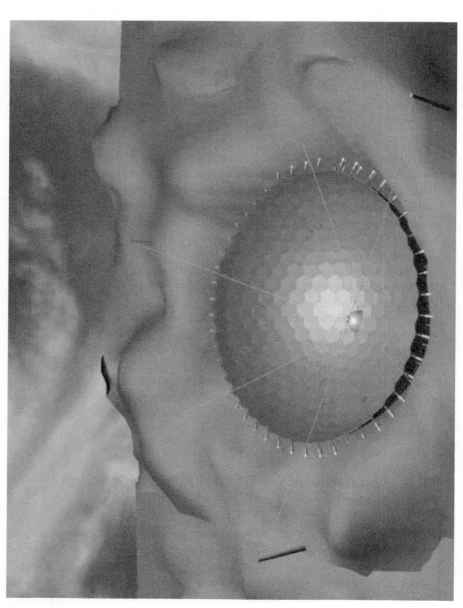

Figure 10.6 Innovative design of the FAST optomechatronics design. Source: Quality Digest Magazine.

the software platform, the innovative design of optical–mechanical–electronic technologies integration has been proposed for the first time in the world (Figures 10.6 and 10.7), which not only reduces the self-weight from 10 000 tons to about 30 tons, greatly reduces the blockage of electromagnetic waves, but also achieves a wide range of dynamic positioning accuracy by mm. At the same time, in combination with the active main reflecting surface, the difficult problem of narrow bandwidth of the line feeder has also be solved [33–37].

To verify the engineering feasibility of the innovative design, three test antennas, FAST 5m, FAST 50m-1, and FAST 50m-2, have been built in Xidian University campus (Figure 10.8), and a large number of experimental results verified the

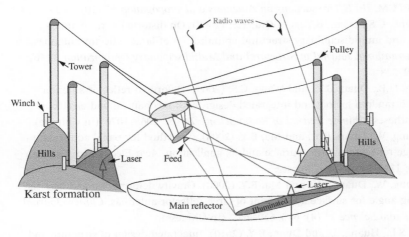

Figure 10.7 Flexible cable support and coarse-finish composite adjustment design for FAST.

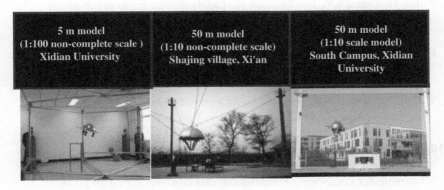

Figure 10.8 Physical photograph of the FAST trilogy of scaled-down model tests.

engineering feasibility of the scheme, clearing key technical obstacles for the implementation of the "Chinese sky eye" FAST500m large radio telescope.

The world's largest single-aperture radio telescope, inaugurated on 25 September 2016, has yielded fruitful scientific outputs, of which more than 500 pulsars alone, including millisecond pulsars, have been observed. The discovery of pulsars is of great significance for the future navigation of human cosmic voyages.

References

1 Duan, B.Y. and Wang, C.S. (2009). Reflector antenna distortion analysis using MEFCM. *IEEE Transactions on Antennas and Propagation* 57 (10): 3409–3413.

2 Wang, C.S., Duan, B.Y., and Qiu, Y.Y. (2007). On distorted surface analysis and multidisciplinary structural optimization of large reflector antennas. *International Journal of Structural and Multidisciplinary Optimization* 33 (6): 519–528.

3 Ma, H.B., Duan, B.Y., and Wang, C.S. (2009). Deformed reflector antenna with random factors and integrated design with mechanical and electronic syntheses. *Chinese Journal of Radio Science* 24 (6): 1065–1070. (in Chinese).

4 Wang, W., Duan, B.Y., and Ma, B.Y. (2008). A method for panel adjustment of large reflector antenna surface and its application. *Acta Electronica Sinica* 36 (6): 1114–1118. (in Chinese).

5 Wang, W., Duan, B.Y., and Ma, B.Y. (2008). Gravity deformation and best rigging angle for surface adjustment of large reflector antennas. *Chinese Journal of Radio Science* 23 (4): 645–650+698. (in Chinese).

6 Li, S.L., Huang, J., and Duan, B.Y. (2010). Integrated design of structure and control for radar antenna servo-mechanism. *Journal of Mechanical Engineering* 46 (19): 140–146. (in Chinese).

7 Bao, H., Duan, B.Y., Du, J.L., and etc. (2008). Simultaneous optimization of control and structural systems of a complex machine. *Chinese Journal of Computational Mechanics* 25 (1): 8–13. (in Chinese).

8 Zhou, J.Z., Duan, B.Y., Huang, J., and etc. (2009). Support vector regression modeling and backstepping control of friction in servo system. *Control Theory & Applications* 26 (12): 1405–1409. (in Chinese).

9 Zhou, J.Z., Duan, B.Y., and Huang, J. (2009). Modeling and effects on open-loop frequency for servo system with backlash. *China Mechanical Engineering* 20 (14): 1721–1725. (in Chinese).

10 Zhou, J.Z., Duan, B.Y., and Huang, J. (2007). Adaptive control of servo systems with uncertainties using self-recurrent wavelet neural networks. *2007 IEEE International Conference on Automation and Logistics, August 18-21*, Jinan, China.

11 Ma, Y.L., Huang, J., and Zhang, D. (2008). Backlash compensation in servo systems based on adaptive backstepping-control. *Control Theory & Applications* 25 (6): 1090–1094. (in Chinese).

12 Ma, Y.L., Huang, J., and Zhang, D. (2009). Adaptive compensation of backlash nonlinearity for servo systems. *Journal of System Simulation* 21 (5): 1498–1504. (in Chinese).

13 Tong, X.F., Huang, J., and Zhang, D.X. (2009). Multidisciplinary joint simulation technology for servo mechanism analysis. *2009 IEEE International Conference on Information and Automation*, June 22–25, 2009, Zhuhai, China.

14 Li, Z.D., Chen, G.D., and Ma, H.B. (2008). Structure factor to radar servo system performance influence and its test research. *Modern Electronics Technique* 31 (3): 33–36. (in Chinese).

15 Zhou, J.Z., Duan, B.Y., Huang, J., and etc.. (2009). Prediction of plane slotted-array antenna electrical performance affected by manufacturing precision. *Journal of University of Electronic Science and Technology of China* 38 (6): 1047–1051. (in Chinese).

16 Yu, W., Gu, W.J., and Guo, X.S. (2008). Pattern analysis for deformed subarray of planar slotted antenna array. *Modern Radar* 30 (12): 70–73. (in Chinese).

17 Sun, Q.F. and Guo, X.S. (2009). Generalized matching equations for slotted-waveguide array antenna impedance match. *Modern Radar* 31 (12): 70–72. (in Chinese).

18 Jia, C.G., Tao, Z.J., Zhang, S., and etc.. (2009). On dual cameras synchronization for high speed photogrammetry system. *Journal of Geomatic Science and Technology* 25 (6): 76–78. (in Chinese).

19 Zhang, K., Luo, F., Li, X.M., and etc. (2009). Random errors analysis of millimeter wave band waveguide slotted linear arrays. *Telecommunication Engineering* 49 (5): 67–70. (in Chinese).

20 He, H.D., Wei, X., Yang, S.P., and etc. (2009). Error analysis of admittance of radiating slot in planar slotted-waveguide array antennas with Monte Carlo method. *Chinese J. Radio Science* 24 (2): 365–368. (in Chinese).

21 Yan, Z.J. (2009). Effect of mechanics structure on the electric performances of radiation element in planar slot antenna. *Telecommunication Engineering* 49 (6): 60–65. (in Chinese).

22 Zou, Y. and Li, X. (2007). A new phased array feed network design. *National Microwave and Millimeter Wave Conference*, 2007.10.18–10.22, Ningbo, Zhejiang. (in Chinese).

23 Xiong, C.W., Hu, G.G., and Wang, Y. (2008). The concept and application of RF conductance profile. *Telecommunication Engineering* 48 (6): 21–24. (in Chinese).

24 Xiong, C.W. and Wang, Y. (2008). Analysis and computation of microwave devices surface RF equivalent conductivity. *Academic Conference on Electromechanical and Microwave Structure Processes*, 2008.9.1~9.5, Jiujiang, Jiangxi. (in Chinese).

25 Zhou, J.Z., Duan, B.Y., and Huang, J. (2010). Influence and tuning of tunable screws for microwave filters using least squares support vector regression. *International Journal of RF and Microwave Computer Aided Engineering* 20 (4): 422–429.

26 Zhou, J.Z., Zhang, F.S., and Huang, J. (2010). Computer-aided tuning of cavity filters using kernel machine learning. *Acta Electronica Sinica* 38 (6): 1274–1279. (in Chinese).

27 Zhou, J.Z., Huang, J., and Ma, H.B. (2010). Modeling the effect of manufacturing precision on electrical performance of filters using support vector regression. *2010 IEEE International Conference on Mechatronics and Automation (IEEE ICMA2010)*. August 4–7, 2010, Xi'an, China.

28 Ma, H.B., Yang, D.W., and Zhou, J.Z. (2010). Improved coupling matrix extracting method for Chebyshev coaxial-cavity filter. *Progress in Electromagnetics Research Symposium*, March 22–26, 2010, Xi'an, China.

29 Shen, Z.F. (2006). Electromechanical coupling analysis of test factors for ESC duplex filters. *The First Academic Meeting of the Research Project on the Basic Problems of Electromechanical Coupling of Electronic Equipment*, 2006.11.10–11.11, Chengdu, Sichuan. (in Chinese)

30 Suo, Y.Q. Research on electromechanical coupling test methods for ESC filters. *The First Academic Meeting of the Research Project on the Basic Problems of Electromechanical Coupling of Electronic Equipment*, 2006.11.10–11.11, Chengdu, Sichuan. (in Chinese)

31 Zhen, L.D. (2009). A study on fabrication of electrically tuned duplex. *Electro-Mechanical Engineering* 25 (5): 46–49. (in Chinese).

32 Shen, Z.F. (2009). ESC duplex filter detection simulation and optimal design. *Defense Manufacturing Technology* 5: 52–55. (in Chinese).

33 Duan, B. (1999). A new design project of the line feed structure for large spherical radio telescope and its nonlinear dynamic analysis. *International Journal of Mechatronics* 9 (1): 53–64.

34 Duan, B. and Qiu, Y. (2001). Modeling, simulation and testing of an optomechatronic design of a large radio telescope. *International Journal of Astrophysics and Space Science* 278 (1-2): 231–239.

35 Duan, B. and Jingli, D. (2007). On analysis and optimization of an active cable-mesh main reflector for a giant Arecibo-type antenna structural system. *IEEE Transactions on Antennas and Propagation* 55 (5): 1222–1229.

36 Duan, B., Qiu, Y., Zhang, F., and Zi, B. (2009). On design and experiment of the feed cable-suspended structure for super antenna. *International Journal of Mechatronics* 19 (4): 503–509.

37 Duan, B. (2018). On innovation, simulation, model experiments and engineering of the five hundred meters aperture spherical radio telescope (FAST), Invited Report at EuCAP'2018, London, United Kingdom, April 9–13, 2018.

11

Development Trends of Electromechanical Coupling Theory and Method of Electronic Equipment

11.1 Introduction

Although the development of electromechanical coupling technology for electronic equipment has made promising progress and brought considerable benefits, its theory, methods, technology, and applications are beginning to emerge, which is not only useful but also full of challenges [1, 2]. In the future, new expectations and demands on electronic equipment, such as extreme frequencies, extreme environments, and extreme power, are demanded, which are described below.

11.2 Extreme Frequencies

Based on the demand for large telecommunication capacity, high transmission rate, and high resolution, the operating frequency of electronic equipment is constantly increasing, which has been improved from microwave, millimeter wave, to submillimeter wave and even terahertz wave. At present, the meter- and centimeter-wave detection radars are relatively mature, and the millimeter wave is widely used in telecommunication systems. The high-speed short-range transmission equipment is being developed at frequencies of up to tens of gigahertz (GHz), and research into the application of higher terahertz is beginning to emerge. For the radio astronomy field, the operating radio waves can be up to hundreds of gigahertz (GHz), such as the world's largest fully movable radio telescope QTT110 m, which is being built in Urumqi, Xinjiang, China (Figure 11.1) and operates at up to 115 GHz with an aperture of 110 m.

It requires the reflector surface accuracy as high as 0.2 mm (rms) and the pointing accuracy as high as 2.5 arc second, which is unimaginably difficult for a huge object with an area of 26 basketball court size, 30-story building height, and 5500 tons weight. For designing such ultrahigh precision, super large

Electromechanical Coupling Theory, Methodology and Applications for High-Performance Microwave Equipment, First Edition. Baoyan Duan and Shuxin Zhang.

Figure 11.1 The world's largest fully steerable QTT110 m radio telescope.

aperture fully steerable dual reflector antenna, the electromechanical separation technology will be no way; even the electromechanical coupling technology will face unprecedented huge challenges. So, it is no doubt that, deepening the electromechanical field coupling theory model and influence mechanism research is imperative, for example the next generation of rain detection space deployable antenna, which is required to work at terahertz, such as 427 THz.

The other is the very low-frequency band, such as meter-wave radar for anti-stealth detection and low-frequency and ultralow-frequency antenna for submarine telecommunication and earth detection. For the above electronic equipment with extremely low- and extremely high-frequency bands, the available electromechanical coupling theory and methods need to be studied deeply and carefully. Generally speaking, at least some improvement is needed to deepen the existing electromechanical coupling technologies, and some brand-new explorations are probably needed.

11.3 Extreme Environments

With the rapid development of science and technology, high-performance electronic equipment will be applied in many regions and boundaries that have never been involved in the past, such as deep space, deep sea, and the three poles of the Earth (North and South Poles, Everest). The working environment of these places is seriously bad, not to mention its own, for example the high and low temperature in space (100° or 200° above and below zero), weight loss, irradiation, ultrahigh pressure in the deep sea, and ultralow temperature in the Earth's three poles. In order to design and manufacture the high-performance electronic equipment and make them work normal that can cope with these harsh environments, the existing electromechanical coupling theory and method may not be effective any more. It is necessary to investigate the new field coupling theory model and the influence mechanism of nonlinear factors on electrical performance.

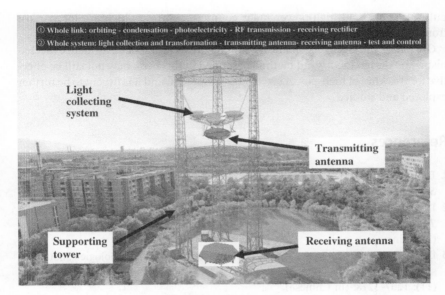

Figure 11.2 Physical picture of ground demonstration and verification system based on continuous microwave wireless power transfer. Source: Xidian University News Network QQ.

11.4 Extreme Power

Another requirement for future electronic equipment is high power. First, continuous high-power microwave wireless transfer technology not only brings unprecedented challenges to the design and manufacturing of transmitting and receiving rectifier antenna but also puts forward new requirements for high-efficiency power amplifier and high-power and high-efficiency rectifier equipment. There are many electromechanical coupling problems in these applications, and many problems are waiting for breakthrough [3]. The other is the space solar power station, which has a large number of multifield coupling theory and method problems waiting for breakthrough from the high magnification concentrated light subsystem, photoelectric conversion subsystem, and microwave conversion and emission subsystem. The OMEGA [4] ground demonstration device is shown in Figure 11.2, which is located in the South Campus of Xidian University, China. Based on continuous high-power microwave wireless transfer, it can simulate the whole link, the whole system wireless power transfer including the light collection, microwave wireless transmitting, and microwave receiving and rectification, which has achieved gratifying stage results [5–8]. Third, the electronic warfare application, such as very high-power microwave pulse weapons and high-power radar countermeasures, which are inseparable from the microwave antennas for very high power.

Obviously, the electromechanical coupling theoretical model, influence mechanism, and design theory and method described in this book may not still be effective for this kind of high-power electronic equipment, and it is urgent to be further explored and studied.

References

1 Duan, B. (2015). Review of electromechanical coupling of electronic equipment. *Scientia Sinica Information*, Science China Press 45 (03): 299–312. (in Chinese).

2 Duan, B. (2021). Electromechanics: being toward electromechanical coupling. *The Introductory Remark of Science & Technology Review* 34 (1): 1–2. (in Chinese).

3 Duan, B. (2018). The main aspects of the theory and key technologies about space solar power satellite. *Scientia Sinica Technologica*, Science China Press 48 (11): 1207–1218. (in Chinese).

4 Yang, Y., Duan, B., Huang, J. et al. (2014). SSPS-OMEGA: a new concentrator system for SSPS. *Chinese Space Science and Technology* 34 (05): 18–23. (in Chinese).

5 Duan, B. (2017). On new development of Space Solar Power Station (SSPS) of China, *National Space Society's 36th International Space Development Conference (ISDC'2017)*, St. Louis, Missouri, USA, May 24–29.

6 Li, X., Duan, B., Song, L. et al. (2017). Study of stepped amplitude distribution taper for microwave power transmission for SSPS. *IEEE Transaction on Antennas and Propagation* 65 (10): 5396–5405.

7 Duan, B. (2019). The Updated SSPS-OMEGA design project and the latest development of China. *Keynote Speech at the 3rd Asia Wireless Power Transfer (AWPT2019)*, Xi'an, China, between Oct. 30 and Nov. 2019.

8 Li, X., Duan, B., and Song, L. (2019). Design of clustered planar arrays for microwave wireless power transmission. *IEEE Transaction on Antennas and Propagation* 67 (1): 606–611.

Index

Electromechanical Coupling Theory, Methodology and Applications for High-Performance Microwave Equipment,
First Edition. Baoyan Duan and Shuxin Zhang.
© 2023 The Institute of Electrical and Electronics Engineers, Inc. Published 2023 by John Wiley & Sons, Inc.

Printed and bound by CPI Group (UK) Ltd, Croydon, CR0 4YY

16/04/2025

14658573-0002